T0325080

Diseases of Commercial Crops and Their Integrated Management

Dr. Amar Bahadur
Assistant Professor
Department of Plant Pathology
College of Agriculture, Tripura
Lembucherra - 799210, West Tripura

Dr. Pranab Dutta
Associate Professor (Plant Pathology)
School of Crop Protection
College of Post-Graduate Studies in Agricultural Sciences
Central Agricultural University
Umiam - 793103, Meghalaya

–EPH–
Elite Publishing House

First published 2024
by CRC Press
4 Park Square, Milton Park, Abingdon, Oxon, OX14 4RN

and by CRC Press
2385 NW Executive Center Drive, Suite 320, Boca Raton FL 33431

CRC Press is an imprint of Informa UK Limited

© 2024 Elite Publishing House

The right of Amar Bahadur and Pranab Dutta to be identified as author(s) of this work has been asserted in accordance with sections 77 and 78 of the Copyright, Designs and Patents Act 1988.

British Library Cataloguing-in-Publication Data
A catalogue record for this book is available from the British Library

Print edition not for sale in India.

ISBN13: 9781032627861 (hbk)
ISBN13: 9781032627885 (pbk)
ISBN13: 9781032627908 (ebk)

DOI: 10.4324/9781032627908

Typeset in Adobe Caslon Pro
by Elite Publishing House, Delhii

–EPH–

Contents

Preface

The edited book on **Diseases of Commercial Crops and Their Integrated Management** made and covered different topics of the national and international economically important crop diseases. An effort has been made to compile information on different aspects of diseases of commercial crops and their management. There is a record of huge losses of commercial crops in terms of yield and quality which are caused by different plant pathogens. It has been felt necessary to compile the information related to the problem of important commercial crops of India and abroad and their integrated management.

This edited book contains 17 crop chapters contributed by different authors throughout the country. The book will helpful for the students pursuing their degree in Agricultural Sciences, growers, teachers, extension personnel, and fellow researchers in their respective fields. This book will also help as a ready reckoner for the readers.

Contribution and cooperation received from the contributing authors of the respective chapter are duly acknowledged. The Editors express their sincere gratitude to all those who helped directly or indirectly in preparing of the book.

The Editors feel great pleasure in expressing a deep sense of gratitude to Elite Publishing House, New Delhi for the publication of this edited book so efficiently and promptly.

Amar Bahadur
Pranab Dutta

About the Editors

Dr. Amar Bahadur is working as Asstt. Professor in the Department of Plant Pathology, College of Agriculture, Tripura, Lembucherra, Agartala-799210. He did his Ph. D. in Mycology and Plant Pathology, at Banaras Hindu University, Varanasi in 2004. He is a Life member of a different society. Dr Bahadur has published more than 33 research papers in International and National Journals, more than 30 proceedings and book chapters in edited books and 3 books. Reviewer of National and International journals. He has been the Principal investigator and Co-investigator in DBT funded research project. He has participated in several seminars/symposia/conferences national and international and also visited Thailand. He was Zonal President of NEZ (North Eastern Zone) in 2018 of Indian Phytopathological Society, IARI, New Delhi. Organizing secretary and organized a National Symposium in 2019. He has engaged in research, teaching and extension activities for more than 15 years. He has awarded the "Outstanding Scientist Award" in 2021 by the VDGOOD Professional Association of India. "Best Senior Scientist" in 2021 by Novel Research Academy, Puducherry, "Best Young Scientist" in International Conference on ICAIR-2021, "Outstanding Faculty" in 9th Faculty Branding Awards-2021 at Kolkata, India, "Out Standing Scientist in Agriculture" in the 2nd International e-conference in 2021 organized by Scientific Educational Research Society Meerut, India. "Excellence in Teaching" in 2021 by SVWS, Lucknow, India, Research Padma Award in 2021 **by ISSN,** Tamil Nadu, India, and Fellows of different academic societies in India.

Dr. Pranab Dutta is an Associate Professor in Plant Pathology in the School of Crop Protection, College of Post Graduate Studies in Agricultural Sciences, Central Agricultural University, Umiam, Meghalaya. He served Assam Agricultural University, Jorhat, Assam, KVK-West Tripura (Now Khowai), CMER &TI (CSB) etc. He completed MSc and Ph.D. in Biological Control and Post Doctorate training at FERA, York, United Kingdom. He has been actively involved in teaching, research and extension for the last two decades. He has published 100 research papers, 45 research abstracts, 13 proceedings paper, 81 popular articles (English, Bangla and Assamese), 22 training manuals/hand books, 36 book chapters, 9 practical manuals /teaching aids, 12 books (Edited, authored or co-edited). Dr. Dutta has planned, participated in and given invitational lectures at numerous national and international conferences and workshops. He is the recipient of the meritorious contribution award in sericulture (CMER&TI, Central Silk Board,) ICPP-2018 Bursary award (The American Phytopathological Society, USA), Young Scientist Award- 2018, Distinguished Scientist Award-2019, Prof. K S. Bilgrami Award-2016, Stanford Research Award-2015, Bharat Siksha Ratan Award, Fellow of FLS (London), FAPPS (New Delhi) and Society for Biotic and Environment Research, Sociery of Biocontrol Advancement, Annals of Plant Protection Society, Linnean Society of London etc. Developed 7 bioformulations with native isolates, filled 5 patents and commercialized five bioformulation products. Pursued 18 externally funded projects funded by DBT, DST, Osceola County, Florida, USA, (through CABI via University of Florida), PCIL, AYUSH, RKVY, BRNS etc. He is the life member of 8 professional societies and reviewer of 7 international and nationally reputed journals published by Elsevier, Springer etc. He is the Editor-In -Chief of the Journal of Plant Health Archives and Editor on Indian Phyotopathology. He organized two ICAR sponsored 10 days short course on Biological control during 2017 & 2019 and a national workshop on Biocontrol as Director. He has supervised 3 Ph.D. and 13 M.Sc. students as Major adviser and more than 100 students as member of the advisory committee. His research area includes nanotechnology, biological control, ecotoxicology and plant defence.

Chapter - 1

Diseases of Tea (*Camellia sinensis*) and their Integrated Management

Abhay K. Pandey[1], Manjunath Hubbali[2] and Azariah Babu[1]

[1]Tea Research Association, North Bengal Regional R & D Center,
Nagrakata-735225, Jalpaiguri, West Bengal,
[2]University of Horticultural Sciences, Bagalkot -587104, Karnataka

Introduction

The tea plant (*Camellia sinensis* (L) O. Kuntze), one of the most popular perennial crops, belongs to the family Theaceae. Tea is the second most consumed beverage worldwide after water. In 2017, the worldwide production of tea was 6.1 million t, and the area under cultivation is around 5 million ha. Tea is widely produced in Asia, Africa, and South America. Recently, the cultivation of green and black tea has newly started in New Zealand. India, China, Kenya, Vietnam, Indonesia and Sri Lanka are the major tea-growing countries with China (annually 2.5 MT) is the major producer (FAO 2017) followed by India (1.3 MT). Processed tea and tea plants are endowed with several medicinal values and are also used as an antioxidant in human disease ailments (Yang *et al.*, 2014). Chemically tea leaves are composed of caffeine (~30-90 mg), and a small amount of theobromine and theophylline (Hicks *et al.*, 1996).

Depending on the fermentation process tea has been categorized into three types. The unfermented form is green tea, a partially fermented oolong tea and fermented teas, i.e., black and red tea. Fermented teas undergo a post-harvested fermentation stage before drying and streaming (Mahmood *et al.*, 2010). Commercial tea cultivars include *C. sinensis* var. *sinensis* (China-type small-leaf variety), *C. sinensis* var. *assamica* (Assam-type large-leaf variety), and *C. sinensis* var. *lasiocalyx* (Cambod-type). Recently, tea products have increasingly become a primary source of income and available for consumption in several countries, exclusively for those in developing countries. However, in tea cultivation, several insect pests and diseases are limiting production.

Tea gardens infected with diseases are recognized by loss in yield and quality of production in the changing climate. Approximately, all the parts of tea bushes such as foliage, stem and roots are susceptible to diseases. On average, 10-30% of annual crop losses have been reported in tea due to diseases (Barthakur 2011). Among stem, foliar and root rot diseases of tea plants, foliar diseases are important because they affect harvestable shoots of two leaves and a bud, resulting in massive direct crop loss which may be as high as 43% (Baby 2001; Chen *et al.,* 2018). Tea plants are challenged by many root rots, stem and foliar fungal diseases, however, worldwide blister blight, grey blight, brown blight, twig dieback/stem canker, charcoal stump rot and red root rot are particularly important, and bird's eye spots and *Fusarium* dieback are emerging problems and need more attention than the other fungal diseases which are specific to locations and have a negligible economic impact. Therefore, the objectives of this chapter were to bring together past and recent available literature from global studies on major disease distribution in tea crops, their impact on tea production, and available management strategies, and make it more readily available to Plant Pathologists and Advisory Officers of tea gardens around the world.

Health benefits of tea

In particular, the major biochemical constituents of tea include polyphenols (20-50% flavanols, Yao *et al,*. 2004), alkaloids (caffeine), amino acids, and vitamins which, in addition to health benefits to the consumers (Khan and Mukhtar 2013), contribute to the quality (Engelhardt 2010) as well as protection against biotic (Punyasiri *et al.*, 2005) and abiotic (Zeng *et al.,* 2018) stresses. Tea, in addition to a beverage, tea leaves and their extracts have been widely used as functional foods, cosmetics, and in materia medica. The health benefit of tea includes the prevention of cancer, diabetes, cardiovascular and neurodegenerative diseases, reduction of body weight, and alleviation of metabolic syndrome (Khan and Mukhtar 2013). Theanine (y-ethylamino-L-glutamic acid), a unique amino acid derived during the fermentation of tea, has been shown to have neuroprotective effects and antiviral activities against influenza, and SARS (Severe acute respiratory syndrome) viruses including the new SARS-CoV-2 (Xu *et al.* 2017; Sharma *et al.* 2020).

Diseases of the tea crop

Tea crops are associated with many root rots and foliar diseases. However, the scope of this chapter is limited to geographical distribution, economic impact and epidemiology of some worldwide economically important diseases causing significant yield loss to tea crops. These include blister blight, grey blight, brown blight, red rust, twig dieback/stem canker, black rot, bird's eye spots, *Fusarium* dieback, violet root rot, brown root rot, charcoal stump rot, black root rot, brown root rot, and red root rot. These diseases have also been grouped under primary (red root rot, charcoal stump rot, black rot, and blister blight) and secondary (bird's eye spot, grey blight, brown blight, and *Fusarium* dieback) categories depending upon their nature of occurrence (Barthakur 2011). Primary diseases occur at any time, while secondary diseases arise generally under poor agronomic practices. A few diseases, such as wood rot and cankers are perennial in nature while some diseases like blister blight, grey blight, dieback, and red rust are seasonal diseases.

Foliar diseases, their impact on tea production and management strategies

1. Red rust - *Cephaleuros parasiticus* Karsten

Geographical distribution and impact

The disease is distributed in India and Bangladesh. In Bangladesh, around 40-70% (Hasan *et al.* 2014) and in India 50% yield loss (Baby *et al.* 2001) has been estimated in red rust-affected gardens. The intensity of red rust disease predicts yields of tea in a linear and inverse way (Huq *et al.*, 2010).

Symptoms

Tea leaves are normally affected by epiphytic algae that grow from thin, red or white discs with lobed margins, with rows of filaments rising from a central point. When the disc matures, it produces red filaments and some of these filaments bear sporangia containing spores. If the discs are lichenized, they often turn white. During the algal bloom, the leaves become variegated (yellow or white) and the cells under the algae discolour. The leaves become hardened when the stem colonizes them and develop longitudinal cracks. Occasionally, red rust can cause severe damage to young tea plants by killing patches of stem tissue, resulting in dieback of the stem (Baby *et al.* 2001).

Biology and Epidemiology

The pathogen is an alga and a member of the chlorophyta or green algae. Bushes of all stages are affected by this pathogen. In epiphytic conditions, the pathogen infects only leaves and seeds; while in parasitic conditions it infects stems and branches. Under adverse soil and climatic conditions, young and old tea plants can be afflicted by red rust. According to Sarmah *et al.* (2016), succulent young leaves and shoots are more susceptible to red rust. A number of factors predispose to this condition, such as poor soil fertility, alkalinity, lack of aeration, hardpans, insufficient or complete lack of shade, droughts, and water-logging (Balasuriya 2008). Spores are transmitted by rain, water splash, and wind-driven rain. Bright sunshine, high temperature and humidity favour the disease development. Red rust is not exclusive to tea and is known to infect other plants, including mango, coffee, citrus, and guava.

Disease management

Balanced dosage of nutrients application, the establishment of good shade, correction of soil pH, removal of water logging conditions and irrigation of drought-prone areas are known to reduce the severity of red rust. For stem red rust, spraying of the affected areas with Copper oxychloride 50% WP, first two rounds at the fortnightly interval and subsequent two rounds at monthly intervals has been found effective. In case of red rust on the leaf, spraying of approved systemic fungicides is recommended (Huq *et al.* 2010). Another fungicide Mancozeb has also been found effective against red rust (Islam and Ali 2011).

2. Blister blight - *Exobasidium vexans* Massee

Geographical distribution and Impact

It is an economically important disease of tea plants, was native to India in 1855 (Venkata 1971) and gradually spread to Japan by 1912 and Vietnam by 1930 and later Sumatra and Java (Indonesia) (De Weille 1959). At present, this disease is a serious threat to tea plants in all the tea -growing countries (Uddin *et al.*, 2005; Barthakur 2011) exclusive of Africa and America. The disease was causing 20-50% yield losses in Sri Lanka (Arulpragasam *et al.* 1987), Indonesia (Gulati *et al.*, 1993), and India (Radhakrishnan and Baby 2004) when conditions were favourable for disease development. In addition, disease infection reduces the production of caffeine in tea leaves, which has a key role in plant defense (Ajay *et al.* 2009).

Symptoms

The tea plant is only the host of this pathogen and off all diseases; blister blight had the highest prevalence globally. The sign and symptoms that appear on leaves are lemon green translucent spots on the first and second leaves of harvestable shoots. Basidiospores of the pathogen are produced on the tips of basidia outside the leaf tissues, giving the blisters on the lower surface a powdery white coating. Soon after the release of basidiospores, blisters start to die off from the center.

Biology and Epidemiology

The pathogen, *Exobasidium vexans* Massee, is an obligate, exobasidiomycete. The disease occurs both in at nursery and at mature stages, and only succulent and young leaves are infected by this pathogen. Extended periods of high humidity (80%), cooled temperature (15-25°C) and long leaf wetness periods favour the germination of basidiospores and disease development. Therefore the disease is prevalent during wet, cold weather, and when tea is grown at higher altitudes such as in the Darjeeling regions of India (Barthakur 2011). The pathogen survives in fallen plant debris and is readily spread by the dispersal of spores by the wind.

Disease management

During the monsoon season, trees should be thinned because shade facilitates blister blight development (Balasooriya 2008). As long as infected leaves are removed before sporulation occurs, which can be accomplished through frequent harvesting (shorter blowing rounds) and/or by hard ploughing (fish leaf) of infected leaves, blister blight is kept at a low level. A bacterium, *Ochrobactrum anthropi* has been found effective against blister blight (Sowndhararajan *et al.* 2012). Among the chemical fungicides, copper formulations such as Copper hydroxide, Copper oxide and Copper oxychloride have been found effective in blister blight management (Arulpragasam *et al.* 1987; Ajay and Baby 2010). Rotation of copper fungicides with systemic fungicides such as Hexaconazole, Propiconazole, and Tebuconazole are also recommended for the management of blister blight (Chandra Mouli and Premkumar 1997).

3. Grey blight - *Pestalotiopsis* species

Geographical distribution and Impact

Though the disease was spread in China, Korea, Kenya, Japan, India and Sri Lanka (Park *et al.* 1996; Takeda 2002; Chen *et al.* 2018), however, the pathogen imposed serious problems in tea gardens of Japan (Takeda 2002), Kenya and India (Joshi *et al.* 2009). Yield losses to the disease in severely affected gardens tolled up to 10-20% in Japan (Horikawa 1986; Keith *et al.*, 2006) and in India (Joshi *et al.* 2009).

Symptoms

The symptoms appear as brown concentric spots in the middle of the leaf which later turn grey with a brown margin, spread through whole leaves and cause damage to young developing shoots. The minute dotted acervuli can be seen in the center of lesions. Usually young and mature leaves are attacked by *Pestalotiopsis* and infection on young shoots results in dieback.

Biology and Epidemiology

The pathogen *Pestalotiopsis* is a saprophytic ascomycete, and two species namely *P. longiseta* (Speg.) K. Dai & Tak. Kobay. and *P. theae* have been reported to infect tea shoots worldwide (Barthakur 2011; Sajeewa *et al.*, 2013; Sinniah *et al.*, 2016). Mechanical practices led to the injury of tea plants favour the disease initiation and development (Joshi *et al.* 2009). The pathogen survives in infected crop debris and soil for many years and high temperature (35°C) and relative humidity (>85%) favour the disease development.

Disease management

Collection and destruction of diseased shoots, the appropriate shade of tea gardens, timely irrigation, and application of a balanced dose of fertilizers are helpful in the management of disease (Barthakur 2011). As far as biological control is concerned, 1-2% concentration of *Trichoderma* biocides has been found effective against this disease (Barman *et al.* 2015; Naglot *et al.* 2015; Kumar and Babu 2019). Grey blight can be chemically managed through the use of carbendazim + mancozeb followed by rotation with systemic fungicides like carbendazim and hexaconazole at 15 days intervals (Sanjay *et al.* 2008; Vidhya *et al.* 2012; Barman *et al.* 2015).

4. Brown blight/anthracnose - *Colletotrichum* species

Geographical distribution and Impact

The disease was challenging in Bangladesh, China, Hawaii, Japan, India, Indonesia, and Sri Lanka. In Japan and China, this disease was most widespread (Yoshida *et al.* 2010; Wang *et al.* 2015), however, in other countries it was causing lesser damage to tea plants. Yield damage and economic losses are not well understood for this disease, but recent studies revealed that it caused significant losses to harvestable shoots (Chen *et al.* 2016).

Symptoms

Symptoms on the leaves appear as small, yellowish green, and diffused spots which later become dark brown, necrotic lesions with characteristic concentric rings. The spots on leaves develop from the margin and spread inward. On the lesions, black fructification has been reported and the spots sometimes spread through whole leaves and affect young leaves/twigs and sometimes caused dieback.

Biology and Epidemiology

Brown blight/anthracnose is caused by several species of the saprophytic, ascomycete genus *Colletotrichum*. For instance, *C. camelliae* in China, Sri Lanka, USA (Wang *et al.* 2015; Liu *et al.* 2015; Orrock *et al.* 2020), *C. theae-sinensis* in Japan (Yoshida and Takeda 2006), and *C. gloeosporioides* (Guo *et al.* 2014), *C. acutatum* (Chen *et al.* 2016) in China were widely spread species. The pathogen has an extremely broad host range, causing anthracnose on a variety of crops including cereals, legumes, fruits and vegetables as well as on several perennial crops (Pandey *et al.* 2018). The viability of spores is controlled by several factors including temperature and relative humidity. The pathogen survives in decomposed plant debris and the disease development is favoured by high humidity (>95%) and optimum temperature (25-30°C) as well as by poor air circulation or prolonged leaf wetness (Chen and Chen 1982).

Disease management

Management of disease involves the removal and destruction of diseased shoots, the use of shade in tea gardens, timely irrigation, and the application of balanced fertilizers. In terms of biological control, 1-2% concentrations of *Trichoderma* biocides have been found effective against this disease. The management of brown blight is chemically accomplished by using carbendazim + mancozeb along with a rotation of systemic fungicides like Carbendazim and Hexaconazole at 15 days intervals.

5. Fusarium dieback - *Fusarium solani* (Mart.) Sacc.

Geographical distribution and Impact

Fusarium dieback is restricted to India and is an emerging disease of tea plants and caused by an ascomycete, soilborne fungus *Fusarium solani* (Mart.) Sacc.). In India, it is prevalent in the tea gardens of Assam and West Bengal states (Sharmah *et al.*, 2017; Kumhar *et al.* 2015). Economic impacts for this disease are not documented, however, the investigation of Kumhar *et al.* (2015) evidenced that this disease will be a serious threat to tea plants in India in coming years due to climate change.

Symptoms

The disease infects mature leaves and tender shoots and causes complete necrosis of harvestable shoots from the tip backwards; hence the disease is called die-back. Poor agronomic practices and mechanization cause disease spread and development in tea gardens (Sanjay *et al.* 2008).

Biology and Epidemiology

The pathogen is necrotrophic and is distributed worldwide and has been known to infect beans, peas, potatoes, and many types of cucurbits (*Zhang et al. 2006*). The spores of the pathogen can persist in the soil for a decade, where its chlamydospores overwinter on plant tissue/seed or as mycelium in the soil and spread when splashed by rain. Rainy weather favours its spread, and dry conditions promote its development.

Disease management

In order to manage the disease, it is important to collect and destroy diseased shoots, provide appropriate shade in tea gardens, apply fertilizers at the right time, and apply a balanced dose of irrigation. It has been found that 1-2% concentrations of *Trichoderma* biocides are effective in managing this disease (Kumar and Babu 2019). The chemical method can be used to control the disease by rotating carbendazim + mancozeb or copper oxychloride at 15 days intervals with systemic fungicides like carbendazim and hexaconazole (Kumhar *et al.* 2015).

6. Bird eye spot - *Cercospora theae* Petch.

Geographical distribution and Impact

Bird eye spot is an emerging disease of tea plants in India, China and Sri Lanka (Balasuriya 2008; Gnanamangai and Ponmurugan 2012) caused by an ascomycete, necrotrophic fungus *Cercospora theae* Petch. The yield loss and extent of damage have not been reported for this disease.

Symptoms

The pathogen generally affects major foliage, especially mother leaves and maintenance foliage, which directly distress the active site of photosynthesis and nutrients supply from actively growing harvestable shoots (Baby 2001). The infected leaves show small yellow eye-like spots that slowly expand up to bigger size, and later the outer portion of spots brown while the center becomes grey-white. At the advanced stage of infection, infected leaves lesions spread throughout the leaves resulting in dried leaves and finally causing whole foliage damage.

Biology and Epidemiology

The genus infects a varied range of crops including cereals, conifers, and legumes and is an important destroyer of trees in the forests ecosystem (Kenneth *et al.* 2001). The disease is more dominant in mature tea than in young tea plantations. The pathogen survives in decayed crop debris or planting materials and warm temperatures (30-35°C), frequent rains and high humidity (>85%) favour the disease development (Balasuriya 2008).

Disease management

For disease management in tea gardens, it is crucial to remove diseased shoots, apply fertilizers and irrigation at the right time, and collect and destroy diseased shoots. Studies have found that *Streptomyces sannanensis* or *Trichoderma harzianum* biocides are effective of controlling the disease (Gnanamangai and Ponmurugan 2012). A chemical method of controlling disease is to rotate carbendazim or thiophanate methyl at 15 days intervals with copper oxychloride (Gnanamangai and Ponmurugan 2012).

7. Black rot - *Corticium theae* Bernard & *C. invisum* Petch.

Geographical distribution and Impact

The disease is distributed in India, Bangladesh, and China (Alam 1999). Black rot of tea is of primary nature and is responsible for the direct reduction of crop yield (Ali 1992). According to Tunstall and Sarmath (1947), a bush exposed to black rot to be left untreated for four seasons contributed to an up to 50% reduction in yield.

Symptoms

The disease attacks the leaves of the maintenance foliage just below the plucking table. Mycelium pads at the point of contact hold infected leaves to the next leaf, and they don't fall off. An irregular pattern of dark brown spots appears on the leaf. Dark brown patches form on the leaf, which eventually covers the leaf and cause it to drop off. It initially has a white powdery appearance on its lower surface before it turns black. The fungus produces a visible cord on the stem. Maintenance foliages are severely damaged.

Biology and Epidemiology

In tea crops, the most destructive disease is black rot, a leaf disease caused by two species of *Corticium*, i.e., *C. theae* and *C. invisum*. Basidiospores are carried by workers. High temperatures and humid conditions accelerate the development of the disease. They germinate when it rains and release hyphae that cause fresh infections. Predisposing factors, such as dense shade, bad drainage, sanitary conditions, and high humidity, are usually considered the cause of this disease's occurrence (Ali 1992).

Disease management

Cultural practices such as removing litter from the bush after pruning followed by an alkaline wash, removal of over-dense shades, and improving the aeration by looping side branches and matidals of bushes are helpful in the management of the disease. Besides, drainage should be improved in poorly drained and infested sections. The spraying of *Trichoderma* biocides (2%) or *Bacillus subtilis* (10%) or aqueous plant extracts (5-10%) to control black rot has been proven successful for tea (Kabir *et al.* 2016). At fortnightly intervals, spray two blanket rounds of copper oxychloride rotation with hexaconazole to cover the entire canopy, particularly the lower surface of maintenance foliage, followed by spot treatment until the disease is completely controlled. The biocides can be applied in lieu of copper oxychloride during both active and dormant phases of the disease.

Stem Diseases, their Impact on Tea Production and Management Strategies

Cankers - *Macrophoma theicola* Siemaszko, & *Phomopsis theae* Petch

Geographical distribution and Impact

In almost all tea-growing areas, canker is the most common stem disease. It is difficult to treat cankers. Initially, they attack small branches and twigs, griddle the edges of the branches and eventually kill the entire bush, reducing the lifespan of attacked plants. Initial research in India, China, Kenya, Japan, Bangladesh and Sri Lanka thought canker was exclusively caused by *Phomopsis theae* Petch, a sordariomycete, parasitic fungus (Keith *et al.* 2006; Ahmad *et al.* 2016; Balasuriya 2008). Other parasitic Dothideomycete species, such as *Macrophoma theicola* Siemaszko, have been detected in Sri Lanka, India, and Bangladesh with new diagnostic techniques (Barthakur 2011; Ahmad *et al.* 2016; Sinniah *et al.* 2017). After moving downward through the plant system, the parasitic fungus destroys the entire bush by attacking stems and twigs, killing branches and poisoning leaves. It has been reported that yield losses are approaching 40-50% due to cankers. Tea gardens are drastically reduced as a result of this disease; in particular, it kills 2-8 year-old bushes.

Symptoms

Tea crops are attacked by two types of cankers, namely *M. theicola* stem and branch canker and *P. theae* collar canker. *Macrophoma theicola* infects plants by entering the stems, killing the bark, and causing small elongated and red patches of rot, which are surrounded by a ring of callus. As the disease progresses downward from the point of infection, the bark of affected plants splits. Eventually, the bare wood becomes covered with a fresh layer of re-generated bark which forms from the edges of splits.

However, *P. theae* cankers usually develop from mechanically caused wounds near collars and branches that are associated with collars. A symptomatic plant showed chlorosis, sprinkling, profuse flowers, defoliation, and eventually wilting and death.

In both cases, attacked bushes show brown or yellow leaves and cankers or lesions on the collar regions. The gradual killing of the barks at or around the site of infection led to canker development.

Biology and Epidemiology

Both pathogens are wound parasites and enter the bush frame through wounds. Environmental conditions and cultural practices influence the incidence and severity of canker diseases. Predisposing environmental factors for canker disease include poor soil conditions, drought, and moisture stress (Balasuriya, 2008). Other predisposing factors are planting in gravelly soil, mulching close to the collar, overapplying fertilizer, and watering surfaces. Wind-borne spores spread the disease by surviving for several weeks on branches left in the field and then spreading by rain across the entire plantation, once the spores are splashed with water in the field. Dry conditions promote disease development, and disease severity increases after pruning operations (Barthakur 2011).

Disease management

In dry periods, adequate thatching and adequate shading are necessary to prevent moisture stress on plants. Disease management requires careful removal of all dead wood following every pruning and the collection and burning of infected wood afterwards. It is advisable to smooth off all cuts and wounds with a sharp knife and then paint these areas with Caustic Wash solution or copper fungicide or Bordeaux mixture or *Trichoderma* paste. Furthermore, the use of systemic fungicides can help to manage the disease (Ponmurugan and Baby 2007; Balasuriya 2008).

Root Diseases and their Impact on Tea Production

1. Poria root rot - *Poria hypobrunnea* Petch

Geographical distribution and Impact

The disease is distributed in India, Indonesia, and Sri Lanka (Balasuriya 2008; Chong and Fazila 2012). There are no estimates of yield loss caused by red root rot, but Barthakur (2011) observed that once the disease spreads to the tea bushes, it causes significant losses and at this stage, disease management becomes very difficult.

Symptoms

Tea bushes can be attacked by the pathogen both as young and older plants, and it spreads through wind-borne spores or free-growing white thread-like rhizomorphs in the soil. In the beginning stages of infection, these rhizomorphs are green. They later turn red and form a network over the root surface. As a result of infection, infected roots display red and white spots that gradually disintegrate into a pulp.

Biology and Epidemiology

Poria hypobrunnea Petch. is a saprophytic fungus that inhabits the soil. Humidity and rainy seasons facilitate the growth of the pathogen within the soil, as does soil contamination (Chong and Fazila 2012). Based on soil moisture levels and environmental conditions that promote basidiospore multiplication, the incidence and severity of disease varied from country to country.

2. Charcoal stump rot - *Ustulina* species

Geographical distribution and Impact

It is prevalent in Sri Lanka, India, Malaysia, and Indonesia (Mishra *et al.* 2014) and can lead to significant losses in tea production if the garden is severely affected.

Symptoms

Symptomatic roots lack visible mycelia on the surface, but after removing the bark patches, white mycelia can be seen. Pathogens produce charcoal-like black brittle fructification on the collars or main stems of diseased bushes. The leaves on attacked tea bushes wither very quickly resulting in the sudden death of the tea bushes.

Biology and Epidemiology

Disease is caused by two species of the root parasitic basidiomycetes, soilborne genus *Ustulina* i.e. *U. zonata* (Lev.) Sacc. (syn. *Kretzschmaria zonata* (Lév.) P.M.D. Martin), and *U. deusta* (**Hoffm.**). Direct root contact with infected materials or wind-borne spores spread the disease. Infected soils and decayed infected plants hold viable spores for one year, and the disease becomes severe during rainy seasons and spreads via lateral root contact.

3. Violet root rot - *Sphaerostibe repens* Berk & Broome

Geographical distribution and Impact

Tea gardens affected by this disease could experience significant losses in tea production. Dispersion of the disease occurs in Sri Lanka, India, and Indonesia.

Symptoms

The leaves turn yellow, droop and become floppy. The leaves are usually still green when they fall off. An unpleasant vinegar smell may be noticed in the affected areas. The roots become ink black or light violet in colour. If the bark is removed, the wood's surface is found to be covered in irregular purple strands of fungus.

Biology and Epidemiology

The pathogen, *Sphaerostibe repens* Berk & Broome survives in the soil. The disease can be found in soils of all types; but is particularly prevalent in stiff clayey soils. The disease can affect plants of any age starting at age 2. The pathogen transmitted disease through drainage water. A lack of aeration, flooding, and faulty drainage are among the conditions that facilitate disease development.

4. Black root rot - *Rosellinia* species

Geographical distribution and Impact

This disease is prevalent in Sri Lanka, India, and Indonesia. It occurs in both young and mature tea plants.

Symptoms

The mycelium of the fungus forms irregular, black, cobwebby bands on the root surface. Grey to black mycelium can also be found on the main stem a few inches above the soil surface. An abundance of star-shaped, white mycelium develops under the bark of the wood. On the surface of and embedded in pathogen-infected wood, there are black points or strands that can be seen.

Biology and Epidemiology

The disease is caused by two sordariomycetes fungi, including *Rosellinia arcuata* Petch. and *R. bunoides* Berk. and Br. Infections are spread by wind-borne spores, direct contact with diseased material, or mycelial cords growing freely in the soil.

5. Brown root rot - *Phellinus noxius* (Corner) G.Cunn

Geographical distribution and Impact

In India, Sri Lanka and Indonesia (Morang, 2013), this disease has been reported in all tea growing areas. Yield loss due to this disease has not been estimated. In the early stages of infection, the roots do not exhibit much decay. In later stages, roots are permeated with yellowish-brown sheets, which give them a honeycomb structure.

Symptoms

Once attacked by the pathogen, the roots of the tea bushes are encrusted with a mass of earth and small stones, difficult to remove by washing or rubbing. The mycelium of the fungus is also cemented to the root. These mycelia initially appear as white to brown woolly masses on the surface of roots, and later turn black/dark brown among the earth and stones encrusted to the roots. Between the bark and the wood, there is usually a thin layer of white or brownish mycelium.

Biology and Epidemiology

There are currently 153 host species listed for the pathogen. Some examples include mahogany, teak, rubber, oil palm, tea, coffee, and cacao as well as a variety of fruit, nut, and ornamental trees. Infection occurs when infected material comes into contact with healthy plants (Satyanarana and Venkataramanan, 1979). Basidiocarps form on infected tree trunks or on sawdust medium in culture. They may be overlapping or single and perennial. The fungus often survives for many years on the remains of an infected host plant and has even been recovered from infected tissue 10 years after the host's death. If soil does not contain host plant debris, the viability of the fungus declines rapidly, with no recovery after five months. The viability of this organism is also reduced by flooding. In newly established plantations, soil debris from infected host plants is the most common source of primary inoculum. The inoculum can also come from seedlings infected in the nursery.

Management of root rot diseases

Root infection is difficult to detect early in its progression unlike foliar and stem diseases. In most cases, root diseases become apparent after bushes expire, making it difficult to cure them. It is the only option available to remove symptomatic and dead plants. Rehabilitating soil prior to replanting is highly recommended, as root pathogens can reside within it. For the management of root diseases, the soil was fumigated with methyl bromide (Venkata Ram and Joseph 1974). In the current situation, the management of root disease in tea has become more challenging due to the phaseout of methyl bromide, mainly due to its adverse effects. Under field conditions, fungicide spray is recommended as a preventative measure (Balasuriya 2008). A variety of biocontrol agents are successfully used for eliminating root diseases, including *Trichoderma* (10%) @ 100 ml/pit (Borthakur 2011). Studies found that 50 ml/plant of *Pseudomonas aeruginosa* has been used to control brown root rot (Morang *et al.* 2012), and 100 ml/plant of *Trichoderma atroviride* has been used to control black root rot (Thoudam and Dutta 2014).

The management of violet root rot involves improving soil accretion through proper drainage and maintaining drains in good working order. Tea bushes should not have soil taken away from their base without proper replacement. It is important to avoid piling soil around the collar of the bushes when deepening the drains.

Disease forecasting in tea crop

A disease forecasting model is being used to reduce the need to use fungicides in the management of blister blight in South India and Sri Lanka. It was suggested in Sri Lanka to postpone copper contact fungicide spraying for a period of five consecutive days until the average sunshine time for the five days preceding the postponement exceeded three hours and 45 minutes (Visser *et al.* 1962). Several studies have found that 10 hours of sunshine in the prior four days, or 20 hours in the previous five days, was adequate to delay spraying (Mulder and de Silva, 1960). Venkata (1971), however, demonstrated that this scheme of delayed spraying did not produce satisfactory results in South India.

It has been demonstrated that disease incidence can be predicted three weeks in advance via a combination of sunshine hours and spore counts (Kerr and de Silva 1969). Java (Indonesia) researchers identified microclimatic variables that predicted blister blight infection rates and severity, including leaf wetness duration, leaf temperature and relative humidity. As part of this process, the weather factors were assumed to be simulative also, and plants, pathogens and other cultural practices were assumed to have no impact on the data to be predicted. It was found to be useful, except in cases where the assumptions were incorrect (Gunadi 2001). India has developed a computer-based forecasting model to predict blister blight incidence (Premkumar *et al.* 2002).

Forecast accuracy depends not only on weather parameters; but also on inoculum levels and the availability of susceptible hosts. Therefore, it is difficult to implement forecasting models on a large scale. Furthermore, disease forecasting systems face several logistical challenges for growers. The effectiveness of fungicide spray warnings depends on how convenient it is to adopt, how it may cost and labour to implement, and how accurate it is. Without overcoming these barriers, it is unlikely that warning systems will be adopted by more companies.

Conclusion

In the present chapter, the geographical distribution and economic impacts of tea diseases are discussed. A number of studies conducted in India have been highlighted, and findings from other countries have been included as appropriate. In order to identify the knowledge gaps and future needs of the global tea industry, attention was paid to recent developments in the field of tea pathology, pathogens epidemiology, and management options. Growing tea is currently being managed using fungicides, host-plant resistance, microbial bio-control agents, botanicals, bio-stimulants, and cultural practices. In northeast India, tea growers use biocides (*Trichoderma* species, *Pseudomonas fluorescence*, and *Bacillus subtilis*) as well as botanicals to manage diseases; however, microbial biocides may develop resistance if regularly utilized. In order for biological methods to be fully utilized in integrated disease management programs, long-term field trials with microbial biocontrol agents are required.

References

Ahmad, I., Mamun, M. N. A., Islam, M. S., Ara, R., Mamdud, M. A. A., and Hoque, A. K. M. R. (2016). Effect of different pruning operations on the incidence and severity of various diseases of tea plant. *J. Biol. Sci.* 24:1-9.

Ajay, D. and Baby, U. I. (2010). Induction of systemic resistance to *Exobasidium vexans* in tea through SAR elicitors. *Phytoparasitica*, 38:53–60.

Ajay, D., Balamurugan, A. and Baby, U. I. (2009). Survival of *Exobasidium vexans*, the incitant of Blister blight disease of tea, during off season. *Int. J. Appl. Agric. Res.* 115:23.

Alam, A.F.M.B. (1999). Profile of Tea Industry in Bangladesh. In Global Advances in Tea Science. Araval Books Int. (P) Ltd. P 9.

Ali, M.A. (1992). Black rot disease of tea. Pamphlet no. 14. Bangladesh Tea Res. Inst. Srimangal. pp. 5-10.

Arulpragasam, P. V., Addaickan, S. and Kulatunga, S. M. (1987). Recent developments in the chemical control of blister blight leaf disease of tea-effectiveness of fungicides. *Sri Lanka J. Tea Sci.* 56:22–34.

Baby, U. I. (2001). Diseases of tea and their management-a review. Pointer Publication, Jaipur, pp.315-27.

Balasuriya, A. (2008). Common diseases of tea and their management. In Handbook *on Tea*. (Tea Research Institute of Sri Lanka, Talawakelle, Sri Lanka), pp.173–209.

Barman, H., Roy, A. and Das, S. K. (2015). Evaluation of plant products and antagonistic microbes against grey blight (*Pestalotiopsis theae)*, a devastating pathogen of tea. *African J. Microbiol. Res.* 9:1263-1267

Barthakur, B. K . (2011). Recent approach of tocklai to plant protection in tea in North -East India. *Sci Cult.* 77:381-384.

Chandra Mouli, B. and Premkumar, R. (1997). 'Comparative evaluation of fungicide spray schedules against blister blight (*Exobasidium vexans* Massee) disease of tea', *Pestol.*, 11: 19–21.

Chen, T. M. and Chen, S. F. (1982), 'Diseases of tea and their control in the people's republic of China', *Plant Dis.*, 66: 961–5.

Chen, Y. J., Tong, H. R., Wei, X. and Yuan, L. Y. (2016). First Report of Brown Blight Disease on *Camellia sinensis* caused by *Colletotrichum acutatum* in China. *Plant Dis.*, 100(1): 227.

Chen, Y., Zeng, L., Shu, Na., Jiang, M., Wang, H., Huang, Y., and Tong, H. (2018). *Pestalotiopsis*-like species causing gray blight disease on *Camellia sinensis* in

China. *Plant Dis.* 102:98-106.

Chong, K. P., and Fazila, M. Y. N. (2012). Potential antagonist organisms against *Poria hypolateritia* of red root disease in tea plantation. *Int. J. Biol.* 14:457-460.

De Weille, G. A. (1959). Blister blight control in its connection with climatic and weather conditions. *Archs. Tea Cultiv.* 20:1-116.

Engelhardt, U. H. (2010). Chemistry of tea. In: *Comprehensive natural products II: Chemistry and biology.*(Eds. L. Mender and H.W. Liu), Elsevier, United Kingdom. pp.1000–1027.

F.A.O. (2017). World tea production in 2017; Crops/World Regions/Production Quantity from pick lists. Food and Agriculture Organization of the United Nations, Statistics Division (FAOSTAT). Archived from the original on 11 April 2020.

Gnanamangai, B. M. and Ponmurugan, P. (2012). Evaluation of various fungicides and microbial based biocontrol agents against bird's eye spot disease of tea plants. *Crop Prot.* 32:111-118.

Gulati, A., Ravindranath, S. D. and Chakrabarty, D. N. (1993). Economic yield losses caused by *Exobasidium vexans* in tea plantations. *Indian Phytopath.* 46:155-159.

Gunadi, R. (2001). Blister blight forecasting model based on the relationship between microclimate and disease parameters'. *M. Sc. Thesis*, Bogor Agricultural University, Indonesia. pp350

Guo, M., Pan, Y. M., Dai, Y. L. and Gao, Z. M. (2014). First report of brown blight disease caused by *Colletotrichum gloeosporioides* on *Camellia sinensis* in Anhui Province, China. *Plant Dis.* 98:284.

Hasan, R., Rahman, M. H., Hussain, A., Muqit, A., Hossain, A., Ali, M., and Islam, M. S. (2014). Influence of topography, plant age and shading on red rust (*Cephaleuros parasiticus* Karst.) disease of tea in sylhet region. *J. Sylhet Agril. Univ.* 1:227-230.

Hicks, M. B., Hsieh, Y. P. and Bell, L. N. (1996). Tea preparation and its influence on methylxanthine concentration. *Food Res. Int.* 29:325-330.

Horikawa, T. (1986). Yield loss of new tea shoots due to grey blight caused by *Pestalotia longiseta* Spegazzini. Bull. Shizuoka Tea Experimental Station 12:1-8.

Huq, M., Ali, M. and Islam, M. S. (2010). Efficacy of muriate of potash and foliar spray with fungtcides to control red rust disease (*Cephaleurous parasiticus*) of tea, *Bangladesh J. Agril. Res.*, 35(2): 273–7.

Islam, M. S., and Ali, M. (2011). Efficacy of Sedomil 72 WP and Recozeb 80 WP in controlling red rust of tea. *Bangladesh J. Agril. Res.* 36: 279-284.

Joshi, S. D., Sanjay, R., Baby, U. I. and Mandal, A. K. A. (2009). Molecular characterization of *Pestalotiopsis* spp. associated with tea (*Camellia sinensis*) in southern India using RAPD and ISSR markers. *Indian J. Biotechnol.* 8:377-83.

Kabir, S. E., Debnath, S., Mazumder, A., Dey, T. and Bera, B. (2016). *In vitro* evaluation of four native *Trichoderma* spp isolates against tea pathogens. *Indian J. Fund. Appl. Life Sci.* 6:1-6.

Keith, L., Ko, W. H. and Sato, D. M. (2006). Identification guide for diseases of tea (*Camellia sinensis*). *Plant Dis.* 33:1-4.

Kenneth, H., Rev, By R. and Westcott, C. (2001). Westcott's plant disease handbook, Boston, Mass.: Kluwer Academic Publishers, pp.245-678.

Kerr, A. and de Silva, R. L. (1969). Epidemiology of tea blister blight (*Exobasidium vexans*). *Tea Quarterly.*, 40:9-18.

Khan, N. and Mukhtar, H. (2013). Tea and Health: Studies in Humans. *Curr. Pharma. Design.*, 19:6141-6147.

Kumhar, K. C. and Babu, A. (2019). Biocontrol potency of *Trichoderma* isolates against tea (*Camellia* sp.) pathogens and their susceptibility towards fungicides. *Int. J. Chem. Stud.* 7:4192-4195.

Kumhar, K. C., Babu, A., Bordoloi, M., Banerjee, P. and Dey, T. (2015). Biological and chemical control of *Fusarium solani*, causing dieback disease of tea *Camellia sinensis* (L): an in vitro study. *Int. J. Curr. Microbiol. Appl. Sci.*, 4:955-963.

Liu, Y. J., Tang, Q. and Fang, L. (2016). First report of *Nigrospora sphaerica* causing leaf blight on *Camellia sinensis* in China. *Plant Dis.*, 100:221.

Mahmood, T., Akhtar, N, and Khan, B. A. (2010). The morphology, characteristics, and medicinal properties of *Camellia sinensis* tea. *J. Med. Plant Res.* 4: 2028-2033.

Mishra, A. K., Morang, P., Deka, M., Nishanth, K. S. and Dileep, K. B. S. (2014). Plant growth promoting rhizobacterial strain-mediated induced systemic resistance in Tea (*Camellia sinensis* (L.) O. Kuntze) through defense-related enzymes against brown root rot and charcoal stump rot. *Appl. Biochem. Biotech.*, 174:506-21.

Morang, P. (2013). Integrated management of brown root rot disease of tea (*Camellia sinensis* (L.) O Kuntze.) under the agroclimatic condition of Barak Valley of Assam. *Ph.D. Thesis*, Assam University, Silchar. India . pp150.

Mulder, D. and de Silva, R. L. (1960). A forecasting system for blister blight based on sunshine records. *Tea Quart.*, 31: 56–67.

Orrock, J. M., Rathinasabapathi, B. and Richter, B. S. (2020). Anthracnose in US Tea: pathogen characterization and susceptibility among six tea accessions. *Plant Dis.*, 104:1055-1059.

Pandey, A. K., Burlakoti, R. R., Kenyon, L. and Nair, R. M. (2018). Perspectives and challenges for sustainable management of fungal diseases of mungbean (*Vigna radiata* (L.) R. Wilczek var. *radiate*): A Review. *Front. Environ. Sci.*, 6:53.

Park, S. K., Park, K. B. and Cha, K. H. C. (1996). Gray blight of tea tree caused by *Pestalotia longiseta. Korean J. Plant Pathol.*, 12:463-5.

Ponmurugan, P. and Baby, U. I. (2007). Evaluation of fungicides and biocontrol agents against *Phomopsis* canker of tea under field conditions. *Australasian Plant Pathol.*, 36:68-72.

Premkumar, R., Sanjay, R. and Ponmurugan, P. (2002). Forecasting of blister blight disease of tea. In: Proc. PLACROSYM XIV, Plantation Crops Research and Development in the New Milllenium, pp.535-540.

Punyasiri, P. A. N., Abeysinghe, S. B. and Kumar, V. (2005). Preformed and induced chemical resistance of tea leaf against *Exobasidium vexans* infection. *J. Chem. Ecol.*, 31:1315-1324.

Radhakrishnan, B. and Baby, U. I. (2004). Economic threshold level for blister blight of tea. *Ind. Phytopath.*, 57:195-196.

Sanjay, R., Ponmurugan, P. and Baby, U. I. (2008). Evaluation of fungicides and biocontrol agents against grey blight disease of tea in the field. *Crop Prot.*, 27: 689-694.

Sarmah, S. R., Baruah, P. K. and Das, S. C. (2017). Practical utilization of botanical extracts andmicrobial in controlling dieback disease of tea (*Camellia sinensis* (L) O. Kuntze) caused by *Fusarium solani* (Mart.) Sacc. *J. Tea Sci. Res.*, 7:11-19.

Satyanarana, G. and Venkataramanan, M. N. (1979), Mycorrhiza from tea and associated with in Assam. In *The 2nd Plantation Crops Symposium (PLACROSYM II)*, Ootacamund, India, pp. 84–8.

Sharma, J., Kumar, Bhardwaj, V., Singh, R., Rajendran, V., Purohit, R., and Kumar, S. (2020). An *in-silico* evaluation of different bioactive molecules of tea for their inhibition potency against non structural protein-15 of SARS-CoV-2. *Food Chem.*, 346:128933.

Sharma, M. and Ghosh, R. (2017). Heat and soil moisture stress differentially impact chickpea plant infection with fungal pathogens. In: Plant tolerance to individual and concurrent stresses(Eds. Kumar, S.M .). Springer, New Delhi. pp.47-57.

Sinniah, G. D., Munasinghe, C. E., Mahadevan, N., Jayasinghe, S. K. and Kulatunga, D. C.M. (2017). Recent incidence of collar canker and dieback of tea (*Camellia sinensis*) caused by *Fusarium solani* species complex in Sri Lanka. *Aus Plant Dis. Notes*, 12:41.

Sinniah, G. D., Wasantha, Kumara, K. L., Karunajeewa, D. G. N. P. and Ranatunga, M. A. B. (2016). Development of an assessment key and techniques for field screening of tea (*Camellia sinensis* L.) cultivars for resistance to blister blight. *Crop Prot.*, 79:143-149.

Sowndhararajan, K., Marimuthu, S. and Manian, S. (2012). Biocontrol potential of phylloplane bacterium *Ochrobactrum anthropi* BMO-111 against blister blight disease of tea. *J. Appl. Microbiol.*, 114:209-218.

Takeda, Y. (2002). Genetic analysis of tea gray blight resistance in tea plants. *Japan Agric. Res. Quarterly.*, 36:143-50.

Thoudam, R., and Dutta, B. K. (2014). Compatibility of *Trichoderma atroviride* with fungicides against blackrot disease of tea: an in vitro study. *J. Int. Acad. Res. Multidisciplinary* 2:25-33.

Tunstall, A.C. and Sarmah, K.C. (1947). Black rot of tea in North East India. Memo. no. 19. Ind. Tea Assoc. pp. 1-26.

Uddin, J. M. Shahiduzzaman, M. and Ahmed, I. (2005). Effect of pruning cycle on the yield of mature tea. *Int. J. Sust. Agric.*, 1:82-85.

Venkada Ram, C. S. (1971). The application of forecasting system in the control of blister blight in tea. *Proc. Indian Natl. Sci. Acad.*, 37:377–83.

Venkata Ram, C. S. and Joseph, C. P. (1974), Development in the occurrence and control of root disease in tea. 1. A new root disease incited by species of *Xylaria*. *UPASI Tea Sci. Dep. Bull.*, 31: 6–10.

Venkata, R. C. S. (1979). *Phomopsis* collar canker-A limiting factor to use of clonal material. In *The 2nd Plantation Crops Symposium (PLACROSYM II)*, Ootacamund, India, pp.146-51.

Visser, T., Shanmuganathan N. and Sabanayagam J. V. (1962). The influence of sunshine and rain on tea blister blight, *Exobasidium vexans* massee, in Ceylon. *Tea Quart.* 33:34–43.

Xu, J., Xu, Z., Zheng, W. (2017). A review of the antiviral role of green tea catechins. *Molecules*. 22:1337.

Yang, C.S., Chen, G. and Wu, Q. (2014). Recent scientific studies of a traditional chinese medicine, tea, on prevention of chronic disease s*entary. Medicine*, 4: 17–23.

Yao, L. H., Jiang, Y. M., Datta, N., Singanusong, R., Liu, X., Duan, J., Raymont, K., Lisle, A. and Xu, Y. (2004). HPLC analyses of flavanols and phenolic acids in the fresh young shoots of tea (*Camellia sinensis*) grown in Australia. *Food Chem.*, 84:253-263.

Yoshida, K. and Takeda, Y. (2006). Evaluation of anthracnose resistance among tea (*Camellia sinensis)* genetic resources by wound-inoculation assay. *Japan Agric. Res. Quarterly*, 40: 379–86.

Yoshida, K., Ogino, A., Yamada, K. and Sonoda, R. (2010). Induction of disease resistance in Tea (*Camellia sinensis* L.) by plant activators. *Japan Agric. Res. Quarterly*, 44:391-8.

Zeng, L., Watanabe, N., and Yang, Z. (2018). Understanding the biosyntheses and stress response mechanisms of aroma compounds in tea (*Camellia sinensis*) to safely and effectively improve tea aroma. *Critical Rev. Food Sci. Nutr.*, 59:2321-2334.

Zhang, N., O Donnell, K., Sutton, D. A., Nalim, F. A., Summerbell, R. C., Padhye, A. A. and Geiser, D. M. (2006). Members of the *Fusarium solani* species complex that cause infections in both humans and plants are common in the environment. *J. Clin. Microbiol.*, 44:2186-2190.

Chapter - 2

Diseases of Coffee (*Coffea spp*) and their Integrated Management

Pranab Dutta, Alinaj Yasin, Madhusmita Mahanta, Arti Kumari, Anwesha Sharma

School of Crop Protection, College of Post Graduate Studies in Agricultural Sciences (CPGSAS), Central agricultural University (Imphal), Umiam, Meghalaya

Introduction

The genus *Coffea* is common habitat in Africa and a number of species are scattered in West, Central and East Africa. A number of diseases are the hurdles along with other factors such as low quality, less caffeine content, and poor yield during production only two species are commercially grown worldwide nowadays, namely *C. canephora* (Robusta) in lowlands and *C. arabica* (Arabica) in highlands. Arabica coffee is cultivated under shade in altitude ranges between 1400 and 1800 m. The province of Kaffa in Ethiopia is the origin of these species which was distributed all over the world during the 15th century by Yemen traders.

At present, few remaining rainforests of the world bear coffee growing as understory shrubs under large diversity of trees and shrubs. Even though in some parts of the world, coffee has maintained its own genetic diversity, natural resources are not free from diseases (Muller *et al.*, 2009). However under natural conditions, coffee continues to uniquely survive all hindrances caused by pathogens and pests. But artificial cultivation of coffee calls for definite management tactics to tackle the loss caused by the pest and pathogens (Walker *et al.*, 2007).

Nursery Diseases

Damping off

Symptoms

The fungus cause pre-emergence and post-emergence damping off mostly affect the seedlings in the nursery. The stem and root tissues show rotting at and below the rhizospheric soil. The infected plant becomes weak and water-soaked falls over at base level and dies.

Causal Organism: *Rhizoctonia bataticola*

Disease cycle

Sclerotia in soil are the primary source of inoculum. Dispersal of sclerotia by rain, wind or soil movements causes secondary spread.

Epidemiology

Cold and wet soil with poor soil drainage increases the incidence of disease. An increase in temperature & humidity and deep planting also favour pathogen growth.

Integrated management

Avoid stagnant water and excess humidity by draining off from the nurseries. The infected seedlings should be burnt and the soil, as well as the other seedlings in the vicinity, should be treated with an appropriate fungicide.

Seedling blight

Symptoms

The cotyledons of the infected seedlings often fail to unfold, and the seedlings' wilts and necrotic lesions appear on the stems. Necrotic lesions may appear on the stalk below the unfolded cotyledons in some cases. Eventually, the seedlings die after the enlargement of these lesions.

Causal organism: *Fusarium stilboides.* This fungus is a seed-borne organism and can be spread by insects and rain.

Epidemiology

Usually, the disease occurs in germination beds and nurseries. The seedling blight disease affects almost all the countries where coffee is grown. During days of high and frequent rainfall accelerates the infection. However, the optimum temperature for the occurrence of this disease is 25°C.

Integrated management

A combination of cultural practices along with chemical methods serves better for the management of this disease. The use of disease-free seeds is a priority as the infection is transmitted easily by seeds. Before sowing the seeds should be treated with Benomyl (1 g of Benomyl 50% WP/kg of seeds).

Root Disease

1. Black rot disease

Also known as root rot disease has been reported to cause infection in up to 19% of coffee trees.

Symptoms

The root rot disease is fairly easy to detect as the leaves wither visibly and the branches decay into dry rot. Black or dark brown decaying leaves, twigs and developing berries are the important characteristic symptoms of the disease (Hu *et al.*, 1999). Mycelial threads can be seen on twigs and petioles along with infected leaves hanging down by means of fungal mycelial strands. The infected berries show drooping besides defoliation. Over the affected areas, the sclerotia of the fungus can be seen.

Causal organism

Two species of ascomycetes *i.e. Roselina bunodes* and *Roselina pepo* have been reported to develop best in the injured collar of the trees.

Epidemiology

The trees that grow at a lower altitude are more vulnerable to rapid decay by root rot as compared to those growing at high altitudes. High humidity coupled with warm temperatures and rain aids in the development of the disease.

Disease cycle

Disease transmission takes place mostly by vegetative mycelium spread through contact with leaves. Plant debris also spreads the disease. Fungal mycelium can survive throughout the year.

Integrated management

Cultural management: Removing the disease parts of the plant serves to control the disease.

Chemical control: Copper-based fungicides should be used to avoid the disease to occur in severe.

2. Agaric Root Rot

Agaric or Hymenomycete root rot is one of the serious reasons for the decline of mature coffee trees. They are prevalent throughout the world, but most commonly occur in cooler areas of the tropics, thus *C. arabica* tends to be more susceptible to this disease.

Symptoms

The initial symptoms include the general decline of the coffee plant with chlorosis and leaf shedding leading to the death of the tree. The characteristic symptom of the disease includes the longitudinal splitting of infected roots, stem or collar region. Creamy-white fan-shaped mycelial sheets appear underneath the bark and brown, flattened cord-like rhizomorphs may also be seen. At the advanced stage of the disease, sporophores develop at the tree base during the wet season. Black rhizomorphs can be seen spreading on the rhizosphere and surrounding soil of infected trees.

Causal Organism: *Clitocybe elegans*

Disease Cycle

The root rot occurs in areas where land has been cleaned previously for planting or shade trees are removed. The pathogen survives on older stumps of trees or roots left in the plantation field. Rhizomorphs arising from these roots spread through the soil and infects adjacent coffee trees.

Epidemiology

The disease appears mainly in cooler tropical regions i.e. at higher altitudes.

Integrated management

The management becomes very difficult once the disease appears, uprooting and burning the tree along with the root system remains the only effective control option. Precautionary measures should be taken to avoid the disease such as good sanitation practices in the field, isolation and removal of infected plant parts. The base of the coffee tree must be sprinkled with fungicides but fumigation is generally not recommended. *Trichoderma harzianum* possesses antagonistic potential against the pathogen and thus can act as a potential control option.

Aerial Disease

1. Canker

Symptoms

The pathogen enters through wounds during pruning or other cultivation practices. The leaves fade in colour and gradually turn from yellowish-green to yellow and dieback. Numerous sized cankers, black or brown in colour are visible from scarification out stems upwards. Severity leads to necrosis of infected tissues followed by total girdling of the trunk. (Malaguti 1952).

Causal Organism: *Ceratocystis fimbriata*

Disease cycle

Wounds are the primary mode of infection progress. The fungus produces a large number of chlamydospores present in the soil causing a secondary infection. The fungus is also transported via agricultural tools.

Epidemiology

Temperature ranging from 18-28°C enhances disease development. In certain areas, abiotic stresses like drought or excess rainfall are extremely favourable. Soil deficient in Boron also favours pathogens to grow.

Integrated management

Using resistant varieties, and disease-free planting materials are some of the cultural tactics to manage this disease. The uprooting and destruction of the infected plants provide better management.

2.   Coffee wilt

The disease has been spread very little to till date after being reported firstly on *Coffea arabica* in Ethiopia in 1957. However, the severity of this disease can't be predicted beforehand and may result in severe losses.

Symptoms

Yellowing; and drying falling of leaves are some of the common symptoms when infected by the pathogen. Sometimes a part of the canopy and branches may dry up leading to the death of the plant after the withering of the whole plant. Sometimes the symptoms are categorized into first-degree and second-degree. The first-degree symptom comprises the common number of afflictions and causes which cannot lead to the confirmation of the diseases. Whereas the second-degree symptoms can aid in the confirmation of the disease due to the production of a specific symptom. The symptoms are visible in the place of development of the black perithecia produced by the pathogen on the trunks and small cracks of the bark. The presence of a long black strip bordered by a healthy area underneath the bark can further aid to examine the presence of this particular disease. However, the death of the coffee plant depends on the physiological condition.

Causal organism: *Fusarium xylariodes* or *Carbuncularia xylariodes*

Epidemiology

Very few studies have been made on the epidemiology of this disease. However, the dry climate has constriction the cultivation of coffee and also accelerates diseases of this kind.

Integrated management

The chemical methods of practice alone have failed many times in the management of this disease. Sanitary measures avoiding injuries, and the use of disease-free cultivation tools are important tactics for successful management.

3. Phloem Necrosis

Symptoms

The major, common symptoms are yellowing, wilting and malformation of phloem tissues. This leads to the death of the plant in 3-6 weeks under severe conditions. Other symptoms like dropping of leaves also take place leaving only a bare top with few young leaves at the top.

Causal Organism: *Phytomonas leptovasorum*

Disease cycle

The pathogen is dispersed by insect vectors mostly heteropteran species. The disease can be also transmitted through root grafts but not leaf grafts.

Epidemiology

High humid areas are prone to disease development. The frequent flooding and improper drainage also favour the disease. The fragility of the tree is more likely to increase incidence.

Integrated management

Clean cultivation is recommended along with the use of systemic fungicide.

Foliage Diseases

1. Coffee rust

One of the most devastating diseases coffee rust causes up to 45% economic losses. The severity of this disease has resulted in the shifting of cultivation to other crops apart from coffee in many pockets of coffee cultivation worldwide.

Symptoms

Infections appear on the foliage of the plants. Initially, lesions usually appear on the lowermost part of the plant and then progress upward in the shrub. The observable symptoms on the upper surface of the leaves are small, pale yellow spots. With the advancement of time, these spots enlargement and increase in size. Lesions tend to concentrate around the margin and develop anywhere on the leaves. These lesions eventually enlarge and produce urediniospores. The undersurface of the leaves is covered with orange masses of urediniospores. The fungal masses can be orange-yellow to red-orange in colour. This sequence of events results in the premature drooping of leaves leaving long expanses of twigs devoid of leaves.

Causal organism: *Hemileia vastatrix*

Epidemiology

Excessive rainfall coupled with high humidity and temperature at 18⁰C-25⁰C aids in disease development.

Disease cycle

Urediniospores germinate only in the presence of free water and high humidity, at least 24 to 48 hours required continuous free moisture, infection usually occurs only during the rainy season through stomata on the underside of the leaf. Secondary cycles of infection occur continuously during favorable weather. Urediniospores initiate infections that develop into lesions that produce more urediniospores and dispersed by both wind and rain, splashing rain is an important for local dispersal and long-range dispersal by wind. Urediniospore also dispersal through thrips, flies, wasps, etc. Pathogen survives as mycelium in the living tissues of the host.

Integrated management

Avoid the plantation of a coffee orchard in endemic areas. Following strict quarantine measures can aid in managing the disease. The affected twigs should be removed by proper pruning and cutting. Spraying of copper-based fungicides before and during the rainy season at intervals of 2-3 weeks is effective to manage the disease. Application of systemic fungicides *i.e.* Carbendazim, Benlate, Hexaconazole, Oxycarboxin @ 0.1 to 0.15%.

2. Blister spot

Symptoms

The blister spot disease is characterized by the appearance of small greasy light green spots that appears on the young leaves of a coffee tree. As the disease progresses, the spots become necrotic and coalesce to cover a major part of the leaf. The necrotic leaves soon fall. In severe disease conditions, the mummified coffee fruit sheds and the affected branch wither (Oliveira *et al.*, 2015).

Causal organism: *Colletotrichum gloeosporioides.*

Disease cycle

The disease is primarily seed-borne. However, the genetic diversity of coffee cultivars plays an important role in the expression of the disease symptoms. The secondary spread of the disease is by the conidia which are generally transmitted by rain splashes. Further, eriophyid mites may help in the dissemination of the pathogen.

Epidemiology

The disease development and dissemination are favoured by warm temperature, rainy weather and high relative humidity. Also, the presence of the vector helps in the quick spread of the disease.

Integrated management

The disease can be prevented by growing resistant cultivars and following field sanitation. Seed treatment with *Trichoderma*-based bio formulations can provide effective management against the disease. Chemical fungicides such as Bevistin, Benlate etc. can also be used for the management of blister spots in coffee.

Other Diseases

1. Sooty blotch mould

Symptoms: The disease is caused mainly due to an increase in aphid or scale infestation. Leaves, berries and shoots get covered with black sooty fungal growth, which hampers the photosynthesis process leading to dead tissues.

Causal Organism: *Capnodium braziliensis*

Disease cycle

The pathogen produces spores blown and is dispersed to honey dew-coated plant surfaces. It can survive on the woody tissues till the next season.

Epidemiology

Increase in aphid infestation and increases honeydew secretion leading to sooty mould formation. Wet conditions are favourable for spore germination although warm and dry conditions increase disease incidence.

Integrated management

Management should be directed to eradicate the colonies of aphids and cochineal insects along with the unwanted hosts in the vicinity of the plantations.

2. Pink disease

The pink disease is of minor importance on a coffee plantation and is not known to cause much damage however, sometimes unexpected surges in the disease can also be observed.

Symptoms

The fungal pathogen covers the bark of branches and stems with a salmon pink-coloured powdery coating which later thickens and develops into crust. Sunken areas develop on the stem and plants die due to girdling. The fungus penetrates the bark and eventually infects the cambial layer leading to splitting or cracking. The infected plant parts wither and ultimately death of the plant occurs.

Causal Organism: *Corticium salmonicolor*

Disease cycle

The pathogen survives on numerous alternate hosts *viz.*, citrus plants and various other woody plants such as tea, cocoa, fig, pear etc.

Epidemiology

The disease is favoured under excessive damp conditions but it can also develop in full sunlight. The disease occurs in a severe form when rainfall exceeds 2000 mm per annum.

Integrated management

To prevent the spread of the disease, infected branches and plant parts should be removed and destroyed. Treatment with cupric fungicides such as Copper oxychloride and Bordeaux mixture is effective in disease management. *Tephrosia candida* is very susceptible to this disease and should be avoided as a temporary shade or cover crop in a coffee plantation.

3. Thread blight

The major thread blights are Kolerago or Black rot, Marasmoid thread and horse hair blight which are caused by fungi belonging to Basidiomycetes. Earlier, all three thread blights were considered of minor importance as they chiefly occurred in neglected coffee orchards or in a damp environment. However, recently they have found affecting non-shaded plantations.

a. **Black rot/ Koleroga:** This disease is considered the second most important disease of coffee in our country. The pathogen mainly attacks Arabica and Robusta coffee cultivars and is reported in coffee-growing states of India such as Kerela, Karnataka and Tamilnadu. In severely infected areas, 10-20% crop loss is recorded for the whole plantation.

Symptoms

The early symptoms develop on leaves, berries and tender shoots. The characteristic symptoms include blackening and rotting of infected plant parts. On developing berries, blackening initiates from the side and gradually spreads in a narrow band. Infected leaves get detached from branches and slimy strands of fungal mycelium hang the detached leaves. The infected leaves, petiole and twigs bear threads of fungal mycelium. Infected leaves and berries dry and interwoven mycelium gives a white web appearance. At the advanced stage, premature shedding of leaves and berries occurs.

Causal Organism: *Pellicularia koleroga*

b. **Marasmoid thread blight**

Symptoms

The symptoms produced are morphologically similar to Koleroga's blight. A sheet of greyish mycelium covers the leaves and twigs and ultimately causes plants to wither and die. The leaves remain hanging from the branches within the mycelial sheet.

Causal Organism: *Marasmius scandes*

c. **Horse hair blight**

Symptoms

The infected plant parts bear black threads of fungal mycelium which are generally capped by carpophores. The infected leaves wither and remain hanging from the enveloped mycelium.

Causal Organism: *Marasmius equicrinis*

Disease cycle

The pathogen survives on infected bushes as sclerotia or dormant mycelium. The primary spread of the disease occurs by contact from leaf and bush through dormant mycelium. Infected leaves get detached and disperse via wind causing the further spread of the disease. Wind-borne basidiospores act as a secondary source of inoculum.

Epidemiology

Continuous rainfall without prolonged dry spells, high relative humidity of 95-100 % and dense overhead shade favour the disease development. During monsoons, frequent mist leads to an outbreak of the disease.

Integrated management

The blend of management practices includes maintenance of proper drainage and optimum circulation of air and sunlight. Before, the onset of the monsoon, thinning of the overhead shade should be done. Pruning of bushes to remove criss-cross branches, dry infected branches, and suckers. Affected leaves and berries bearing mycelial thread should be removed before the onset of the monsoon. The curative control measure includes spraying with a 1.0% Bordeaux mixture on both surfaces of the leaves.

4. Dieback

Symptoms: As the name suggests, the disease manifests as drying of the affected branches from tips downward. The leaves and fruit shedding are common symptoms associated with the dieback disease of coffee.

Causal organism: *Colletotrichum coffeanum*

Epidemiology

The disease is caused by the pathogen when the coffee tree is under physiological stress such as inappropriate cultural practices, overproduction etc. a weak tree is more prone to dieback disease than a healthy coffee tree.

Integrated management

The disease can be managed effectively by following healthy cultivation practices. Use of green manure, vermicompost etc. is also suggested. Adequate fertilization, proper irrigation practices, clean intercultural operations etc. can reduce the incidence of dieback disease. The use of biocontrol agents such as *Trichoderma, Bacillus* etc. can manage the disease effectively along with promoting the plant growth parameters.

5. Coffee Berry disease

The coffee berry disease was first identified and reported by Mcdonald's in the year 1922 in Kenya.

Symptoms: The disease is characterized by the development of scabs, i.e., the appearance of irregular-shaped rusty concave spots on the matured berries. As the disease advances actively, more darker and depressed spots develop on the young berries, where a pink-coloured conidial mass can be seen. The acervuli appear as black pinpoints on the depressed lesion in concentric circles. The disease spreads deep inside and damages the pulp and beans making the berries empty and hollow. The berries later dry, mummifies and sheds, thereby severely reducing yield.

Causal organism: *Colletotrichum kahawae*

Disease cycle

The primary infection is caused by the fungus present in crop debris and in soil. The secondary infection is caused by the conidia that are produced on the mummified coffee berries which are carried to the healthy plants by rain splashes.

Epidemiology

The coffee plantations situated at higher altitude (1100-1500m) is more susceptible to the CBD. At higher altitudes, the growth and development of berries are slower than that of low altitude areas, therefore, the pathogen is benefited from the luxury of time to penetrate and cause infection to the berries. Rainfall and fog favour disease development and spread.

Integrated management

Destroying and burning infected berries can prevent the inoculum build-up in the plantation. Field hygiene and adequate intercultural operations are essential for a healthy plantation. *Trichoderma*-based bioformulation can be used for spraying at the time of flowering to protect the berries from probable infection. The spray should be repeated at 15-20 days intervals. The use of copper fungicides such as Bordeaux mixture, Copper oxychloride etc. in a similar manner is also effective in managing the disease.

6. Elgon dieback or bacterial blight of coffee

It is primarily a nursery disease and attacks the younger tissues of the plant.

Symptoms

The symptoms appear as water-soaked leaves that gradually curl up, blacken, wither and die. The pathogen attacks the terminal buds of the tree and spread downward, thereby causing twig die-back. One of the characteristic symptoms of Elgon dieback is the appearance of a cluster of blackened and dried leaves at the tip of the twig, which differentiates it from the twig dieback from overbearing. The coffee flowers and the petioles infected by the pathogen appear black in colour. In severe disease conditions, the whole plant may exhibit a burned appearance (Lopez-Bautista *et al,,* 2007).

Causal organism: *Pseudomonas syringae* pv. *garcae*

Disease cycle

The pathogen can live as an epiphyte on all the aerial parts of the plant till the arrival of favourable conditions to cause infection.

Epidemiology

High rainfall, high relative humidity and cold winds favour the pathogen to penetrate and establish infection within the plant. The coffee plantation in higher altitude areas is more susceptible to this disease.

Integrated management

The disease incidence can be reduced by installing windbreaks at the nursery. The use of clean and sterilized implements for pruning and trimming coffee plants can ensure safety against the probable spread of the pathogen. The use of copper oxychloride is most effective against the disease.

References

Hu, F. P., Ke, Q. H., Tu, R. and Huang, Y. Y. (1999). Causes of Black rot and bitter root of sweet potato in Liancheng, Fujian. *J. Fujian Agri. Uni.*, 28: 441-444.

Lopez-Bautista, J., Rindi, F. and Casamatta, D. (2007). The systematics of subaerial algae. In: Algae and cyanobacteria in the extreme environment. Springer, Netherlands. pp. 599-617.

Malaguti, G. (1952). *Ceratostomella fimbriata* en el cacao de Venezuala. *Acta Cientificia deVenezuala*, 3: 94-97.

Muller R. A., Berry D., Avelino J., Bieysse D. (2009). Coffee diseases. In : Wintgens Jean Nicolas (ed.). Coffee: growing, processing, sustainable producton : A guidebook for growers, processors, traders, and researchers. Weinheim : Wiley-VCH, p. 495-549

Oliveira, F., M.L., Abreu, M.S. and Silva, J.L. (2015). Diagrammatic scale for blister spot in leaves of the coffee tree. *African J. Agric. Res.*, **10**(19): 2068-2075.

Walker, J. M., Bigger, M. and Hillocks, R. J. (2007). Coffee pests, diseases and their management, CAB books, Cabi International Series. pp. 248.

Chapter - 3

Diseases of Cocoa (*Theobroma cacao*) and their Integrated Management

Amal Jyoti Debnath[1], Pranab Dutta[2] and Amar Bahadur[3]

[1]Department of Plant Pathology, College of Agriculture, Assam Agricultural University, Jorhat- 785013
[2]School of Crop Protection, College of Post Graduate Studies in Agricultural Sciences, Central Agricultural University (Imphal), Umiam, Meghalaya-793103
[3]Department of Plant Pathology. College of Agriculture, Tripura, Lembucherra, Agartala-799210

Cocoa or Cacao is a tropical evergreen fruiting tree from the Malvaceae family. Its botanical name is *Theobroma cacao*. The scientific name is derived from two Greek words, namely "Theos" which means "god" and "broma" which means food. Hence, it is also known as the food of God. The tree is native to the Andes in South America (Cheesman, 1944) and it grows in a limited geographical zone of 20° North and South of the equator. The seeds of the cocoa fruit or Cocoa bean which is in the number of 20-60 are arranged in regular rows covered by mucilaginous acidic pulp rich in glucose, fructose and is the most economically valuable parts of the plant which is used as raw materials for the chocolate and chocolate confection production. The cocoa bean is also used for beverages and as a flavouring ingredient after roasting and grinding.

The cocoa solid or powder is the remnant after the extraction of cocoa butter from the bean through processing. Cocoa butter represents 46-57% of the weight of the cocoa bean which gives characteristics melting properties to the chocolate. Cocoa powder has a high nutritive value and 100g of its powder contains 57.90g of Carbohydrate, 13.70g of Fats, and 19.60g of protein including various minerals such as Calcium, Iron, Magnesium, Manganese, Phosphorus, Potassium, Sodium and Zinc. Cocoa is also taken as a beverage and it acts as a stimulant due to the presence of a high amount of alkaloids particularly Caffeine (230mg/100g of cocoa powder) and Theobromine.

Globally 5.2 million cocoa beans had been produced in 2017. About 70% of the worldwide production of cacao beans comes from West African countries like Ivory Coast, Nigeria, Ghana and Cameroon. In India it is mainly cultivated in Kerala, Karnataka, Tamil Nadu and Andhra Pradesh covering a total area of 97563 ha with a production of 25783MT in the year 2019-20 (Directorate of Cashewnut & Cocoa Development).

There are two major recognizable botanical groups of Cocoa and they are Forastero and Criollo. There is a third variety known as Trinitario which is developed due to natural hybridization among the former two groups. The Forastero covers 85% of the total cocoa production due to its disease resistance capacity and easily cultivable but is of low aromatic and poor quality hence generally used as mixture with other valuable varieties. The variety Criollo has a very pleasant and penetrating aroma, and optimum taste and is considered of excellent quality, but its production is very less representing only 5% of the worldwide production due to high susceptibility to various diseases (Bailey and Meinhardt 2016; Delgado-Ospina *et al.*, 2021). Hence appropriate management of the diseases is very much essential to maximize the quality production of all groups of cocoa for meeting the growing demand (Bowers *et al.*, 2001).

Diseases of Cocoa

Fungal Diseases

1. Phytophthora Diseases

a. Black Pod or Phytophthora pod rot

Symptoms

Initially, small translucent spots developed which later turns into dark hard spots. Within 14 days of disease initiation, the whole pod becomes black or brown and necrotic. White to yellow mycelial growth developed on the darkened area of the lesion. In the later stage of the disease development, the internal tissue of the fruit including the bean becomes dry and shrivelled which results in a mummified pod. The pathogen also causes canker on flower cushions, barks and on main roots near the soil line (Vanegtern *et al.*, 2015).

b. Seedling die-back

Symptoms: Infection is severe mostly in 1-4 months old seedlings. Infection may initiate from the collar region, cotyledonary stalk or from the tip. Water-soaked, dark brown to black linear lesion appears which extends through the petioles to the leaf resulting in wilting and defoliation of the seedling. Dieback of the seedling appears in the advanced stage.

c. Stem canker

Symptoms: Initially water-soaked lesion with a dark margin appears on the bark. Oozing out of reddish brown liquid from the lesion is observed which later on dries up to form rusty deposits. Lesions coalesce and cause extensive rotting. The infection spreads up to vascular tissue. Dieback occurs as soon as the canker girdles the stem. Leaves yellowed, wilt and defoliated.

Causal organism: *Phytophthora palmivora, P. megakarya* and *P. capsici*

Disease cycle

The fungus survives as mycelium, chlamydospore or oospores in infected plant parts, canker, soil, and mummified pods or on fallen litter during the dry season. The primary spread of the disease is due to the rain splash which helps the inoculum to reach the target site. The pathogen can attack all parts of the plant at any stage. Under the humid condition, sporangia are produced on the affected plant parts and act as a source of secondary inoculum. These are dispersed through windborne rainfall, water splash and running water. The sporangia can directly germinate on contact with the plant surface or produce zoospores in free water which actively swim towards the host and cause infection on it (Gregory and Maddison 1981).

Epidemiology

Wet weather (temperature around 18-25°C), high humidity, poor drainage heavy canopy and low-hanging branches led to the severe spread of the disease.

Management

The disease can be managed by the removal and destruction of infected plants. Scrape out the surface of the canker and swab it with 10% Bordeaux paste. Planting with proper spacing and weeding for air circulation. Application of copper and metalaxyl-based fungicide such as copper oxychloride @0.25% or Bordeaux mixture

@1% or Ridomil Plus @ 0.3% to the plants during the onset of monsoon and fortnightly during the peak period of incidence. Spraying of *Pseudomonas fluroescens* (pf1) liquid formulation @ 0.5% as foliar or soil application or *Trichoderma virens* liquid formulation @10^6 conidia per ml thrice per year *i.e.* during June, October and February to the soil along with FYM (Costa *et al.*, 1996; Sriwati *et al.*, 2019). Seed dressing and soil drenching with Kocide fungicide @0.91kg in 45 litres of water.

2. Witches' Broom

Symptoms

The fungus attack only on the growing parts of the plant. Colonization of the fungus on the growing tip led to physiological and hormonal changes. This change led to the swelling and development of a number of succulent branches also known as brooms. The brooms die after several weeks and as the weather becomes humid, a pinkish mushroom i.e. the fruiting body of the fungus developed on it. Infection on pods during the earlier stage causes deformation, uneven ripening and necrotic lesion on it preventing the development of the seed. Star broom or abnormally large flower formation, which does not yield any pod.

Causal organism: *Crinipellis perniciosa*

Epidemiology

Formation and maturation of basidiocarp as well as release of basidiospore is severe at 20-25°C with a relative humidity of 80% or more. Germination of basidiospore is highest at 30-33°C on the availability of free moisture on the host surface to cause rapid infection.

Disease cycle

The fungus exhibits homothallism and lives both as biotrophs and necrotrophs. The single-celled basidiospore released from the pinkish mushroom (basidiocarp) is spread through wind and rain onto the host. The basidiospore germinates and penetrates meristematic tissue through epidermis, stomata or trichomes and colonizes there as a biotroph. On the prevalence of congenial environment, Dikaryotic mycelium develops basidiocarp on dead host tissue during its necrotrophic phase thus completing the life cycle.

Management

Removal of diseased parts and spraying of copper fungicide give good disease control. *Trichoderma stromaticum* is very much efficient and able to reduce 99% of basidiocarp formation. 40g of its spore formulation application is enough for 1ha of the cocoa plantation.

3. Frosty pod rot

Symptoms: Initially discoloured and swelled area on pod followed by the development of large dark-brown necrotic spots with irregular borders. Then these necrotic areas are covered by thick frosty white mycelial growth where heavy sporulation of cream-coloured spores developed hence called a frosty pod. Then the necrotic fruit becomes hard and mummified (Batema *et al.*, 2005).

Causal organism: *Moniliophthora roreri*

Disease cycle

It is a hemibiotrophic fungus with a long biotrophic phase. Conidia are produced on the necrotrophic lesion of the pod and are the only known infective propagules which spread to another pod through rain, wind and human activity. They persist as stroma in a necrotic pod.

Epidemiology

Hot rainy weather with a temperature ranging between 22-32°C favours the disease incidence.

Management

Since the pathogen infects only the pod, hence periodic removal of the disease pod before sporulation gives the best result. Regular pruning limits humidity to prevent spore germination. Application of systemic fungicide like Flutolanil (Moncut 50% WP) 300 g *a.i.* per ha and prophylactic application of Copper hydroxide (Kocide 50% WP) @ 1500 g *a.i.* per ha gives suitable results. Apply *Trichoderma asperellum* formulation at a concentration of 5 ×10^{12} conidia per ml to the soil.

4. Vascular streak dieback of cocoa

Causal organism: *Oncobasidium theobromae*

Symptom

Initial symptom is the characteristic chlorosis of one or two leaves usually on the second or third flush behind the growing tip followed by marginal necrosis and shedding of affected leaves resulting in shoot dieback. Three darkened vascular traces were found on the leaf abscission point. Enlarged lenticel and roughening of bark. Dark stripes on the xylem tissue are seen when the stem is split open. The proliferation of lateral buds initially and died later on to give broomstick appearance. During wet weather, white basidiocarp develops on fallen leaves (Guest and Keane, 2007).

Disease cycle

The fungus remains asymptomatically for a period of 3-5 months during which it spread within the plant through the xylem. They remain as mycelium in the infected leaves and stems. On prolonged wet weather prevalence, hyphae emerge from the crack developed on the main veins of infected leaves or from the leaf scars on the stem to give rise to basidiocarp. The basidiospores are forcibly discharged from it which reach new infection sites through the wind. On availability of free water, they germinate on leaves and cause infection (Guest and Keane, 2018).

Epidemiology

Unpruned canopy, heavy shade increases leaf wetness and canopy humidity which results in heavy sporulation and disease spread.

Management

Planting of disease-free material. Pruning diseases stem below 30-40cm of the visible streak and canopy management for proper aeration to prevent sporulation. Apply 1% Bordeaux mixture to the pruned area and spray 0.25% Copper oxychloride solution or 1% Bordeaux mixture twice during May-June and October. Drench the soil with 0.25% Copper oxychloride solution. Some other fungicides are Hexaconaczole EC 50 @ 2.35mg a.i. per plant and Triadimenol EC50 @ 4.17mg *a.i.* per plant (Holderness, 1990).

5. White thread blight disease

Causal organism: *Marasmiellus scandens* and *M. byssicola*

Symptoms

White mycelial growth on leaves, petioles and branches. Leaves blighted and defoliated. The defoliated leaves cling to each other and to the twigs of the tree with white mycelium strands. Extensive death of young branches covered with mycelial growth is seen in the disease field (Amoako-Attah *et al.*, 2016).

Disease cycle

Dead leaves covered with mycelium acts as a source of inoculum which disseminates through wind, rain, nesting bird and human activities causing the spread of the disease. Sometimes white to pale yellow basidiocarp develops on infected leaves or stems which produces airborne basidiospore during the rainy season.

Epidemiology

High humidity and shade, less aeration and sunlight are predisposing factors for the disease occurrence.

Management

Regular pruning and destruction of infected leaves, and stems give the best result. Excess shade reduction and removal of overhanging branches reduce the humidity required for the disease's spread. Chemical control by the spraying of Ridomil gold or Metalm @ 3.3g/l or Nordox @ 5g/l gives good results.

6. Charcoal pod rot

Causal organism: *Botryodiplodia theobroma / Lasiodipoldia theobromae*

Symptoms

The disease generally occurs on wounded pods during the dry season as a yellow colour spot which later turns to chocolate or dark brown in colour. The entire pod surface becomes charcoal black in colour covered by black fungal spores. Disease at the initial stage of pod development led to the underdeveloped bean. The infected pod was mummified and adhered to the plant (Guest 2007).

Disease cycle

During unfavourable conditions, fungus survives as pycnidia. The pycnidia release conidia which get disseminated through rain splash or wind and disease develops as soon as the conidia land on wounded pods or twigs.

Epidemiology

Hot and dry conditions along with stressed plants are predisposing factors for disease development.

Management

Destroy infected pod and follow proper spacing. Prune the tree regularly for aeration and shade management. Do not intercrop with Areca and avoid wounding of pods during farm operation. Apply a recommended dose of fertilizer and provide an irrigation facility. Spray 1% Bordeaux mixture to control the disease. Apply *Trichoderma* viridae liquid formulation @10^6 conidia per ml to the soil along with farm yard manure.

7. *Colletotrichum* Diseases

a. Leaf blight

Symptoms

The initial symptom is seen as a sunken spot which is surrounded by a yellow halo. Necrosis of the central portion of the leaf leads to a drop of the region producing a shot hole. Irregular spots occur on leaves followed by blighting is seen in advanced cases.

b. Pod rot

Symptom

The dark brown discoloured area is surrounded by a yellow halo of young fruits. The disease mainly starts from the stalk end of the pod. At the advanced stage, the whole pod turns brown to black, is mummified and remains in the tree. Sometimes sunken lesions with brown colour may be found on the pod.

Causal organism: *Colletotrichum gloeosporioides*

Disease cycle

The fungus overwinters as mycelium or sclerotia in plant debris or in the soil act as the primary source of inoculum. Conidia are produced abundantly on acervulus in infected plant parts during high humid weather and are disseminated through dew and rain splash or by the wind to new host sites (Mohanan *et al.*, 1989).

Epidemiology

The constant low temperature (20-25⁰C) and prolonged leaf wetness period during monsoon season facilitate germination, infection and high sporulation of the pathogen causing high disease incidence.

Management

Pruning and removal of infected parts, reduction of excess shade allowing proper aeration and following phytosanitary measures. Spraying of fungicides such as 1% Bordeaux mixture or 1% Propiconazole as a foliar spray before onset and during the monsoon period controls the disease spread.

8. Pink Disease

Causal organism: *Erythricium salmonicolor*

Symptom

The fungus mainly attacks branches of cocoa and initially fine white threads of the pathogen are seen on it, which at a later stage covers the branches with brown, orange or pink incrustation. Dieback of such branches with brown dead leaves clinging to them for several weeks is observed. Affected branches also show cracks on bark, dried flowers and mummified pods (Akrofi *et al.*, 2014).

Disease cycle

The fungus can survive in the collateral host or overwinters in disease brunch and trunk canker. The disease spread through spores produced on the crust and is disseminated by wind or rain to a new host site.

Epidemiology

Since the spores need free water for germination hence rainy season is most favourable for disease incidence with a temperature around 28^0C.

Management

Removal and destruction of infected part, shade reduction, providing good drainage, removal of nearby collateral hosts such as tea, orange, mango, citrus etc. and spraying of copper-based fungicide such as Kocide 2000DF (Copper Hydroxide 53.8%) @100g per 15 litres of water at 2weeks interval during monsoon season effectively control the disease.

9. Ceratocystis wilt of cocoa

Causal organism: *Ceratocystis cacaofunesta*

Symptom

Disease incidence is indicated by drooping of matured leaves followed by wilting, and drying and the dead leaves remain attached to the dead twigs or branches for several weeks. Wilting and rapid death of plants is observed at the advanced stage. Discolouration in infected vascular bundles can also be observed. Canker is produced at a later stage of infection on mature plants (Engelbrecht *et al.*, 2007).

Disease cycle

The fungus penetrates the tree and reaches the xylem through wounds where the chlamydospore germinates and invades host tissue and causes necrosis and wilt. It is homothallic and reproduces asexually through conidia or vegetative propagation and sexually by producing ascospore. These spores are disseminated to new sites through rain splash and wind. Nitidulid and ambrosia beetle which feed on fungal growth act as a vector for ascospore dissemination. The fungus overwinters as mycelium within the host or as chlamydospores in soil or plant debris.

Epidemiology

The fungus grows profusely in a temperature range of 18-28°C and produces ascospore profusely within a week. Wounds created during harvesting, pruning or by borer provide the site for infection to the pathogen.

Management

Use disease-free transplanting material; avoid injury to the plant during harvesting or pruning. Application of 20 ml of 14.3% propiconazole to the collar region through injection reduces disease development.

10. Verticillium wilt of Cocoa

Causal organism: *Verticillium dahliae*

Symptom

The disease is characterized by sudden wilting followed by necrosis of leaves. Brown colour discolouration on vascular tissue, tyloses and deposition of gums, gel on it appears.

Disease cycle

It is a root inhabiting fungus and overwinters as microsclerotia. These microsclerotia germination is stimulated by root exudates which penetrate through the root tissue and produce conidia abundantly. The conidia move to the aerial parts through the xylem transpiration stream and colonize there resulting in necrosis of tissue. Microsclerotia are produced abundantly in these necrotic tissues. After the plant dies, these structures got incorporated into the soil or in the plant debris.

Epidemiology

Moist soil with a temperature range of 25-28^0 C favoured the pathogen growth. The incidence is more in the presence of alternate hosts such as crops from the family Malvaceae, Solanaceae, Compositae etc. which helps in the inoculum buildup of the fungus.

Management

Soil solarization, planting of healthy transplanting material, removal of the infected crop, adequate irrigation, balanced fertilization and application reduce disease incidence.

Viral Diseases

11. Cocoa swollen shoots virus disease

Symptoms

Swelling of nodes, internodes, meristematic tip and tap root is observed. Sometimes necrosis of fibre roots is observed. Leaf symptoms include banding and reddening of the primary vein in young leaves. Vein clearing shows a fern-like pattern with chlorosis and mottling in the case of matured leaves. The abnormally shaped pod is usually small and spherical.

Causal organism: Cocoa swollen shoot virus

Disease cycle

The disease is transmitted by 14 species of the mealybug through the semi-persistent mode of transmission. There are reports of seed transmission of the disease. It is also transmitted mechanically or by grafting. Some of the alternate hosts are forest tree-like *Cola chlamydantha*, *Sterculia tragacantha* and *Erytropsis barteri*

Vector

Mealy bugs mainly *Planococcus citri*, *Planococcoides njalensis* and *Ferrisia virgata*

Epidemiology

Since the disease is spread through the Mealy bug hence the environment favourable for the insect increases the disease incidence. Warm weather of 25⁰ C with high humidity along with high-density planting in a windy environment helps in easy dispersal of the vectors to neighbouring plants.

Management

Removal, uprooting and destruction of the infected tree along with adjacent plants. Planting of disease-free material, apply only recommended dose of fertilizer. Spot application of neem seed kernel extract or Neem oil @5% with 1%detergent solution or application of insecticide such as Profenphos @1.5ml/l with 1% soap solution. Spraying of quinalphos 25EC @1g/l or quinalphos 25 EC @2ml/l to control the vector.

12. Cocoa necrosis virus disease

Causal organism: Cacao necrosis virus

Symptom

Chlorotic spot, midrib and main vein necrosis of leaves at early stage followed by terminal dieback of shoots at a later stage of infection.

Disease cycle

The disease is transmitted through grafting with diseased rootstock. Few reports say that the disease is possibly transmitted through *Longidorue* spp. of nematode.

Epidemiology

Condition favourable for nematode such as loose moistened sandy soil with mild temperature around 25⁰C.

Management

Transplant disease-free material and removal and destruction of the infected plant.

Root Knot Nematode disease

Causal organism: *Meloidogyne* spp. (Prominent)

Symptom

Stunted and reduced plant growth, small leaves eventually yellowed and dried. Formation of galls swelling of roots, dieback of the shoot and wilting are commonly observed.

Disease cycle

Disease spread through infested seedlings and soil and through runoff water. Since the pathogen has a wide range of hosts, hence they can also act as the primary source of inoculum.

Management

The movement of planting material and water from the infested area should be prohibited. Crop rotation with non-host crops and soil solarization will reduce the nematode population. The seedling should be obtained after proper examination. Soil drenching with Fenamiphos 40EC (Nemafose) @3l/acre and Oxamyl 10EC (Vaydet) @20kg/acre gives good control.

References

Akrofi, A. Y., Amoako-Atta, I., Assuah, M., & Kumi-Asare, E. (2014). Pink disease caused by *Erythricium salmonicolor* (Berk. & Broome) Burdsall: An epidemiological assessment of its potential effect on cocoa production in Ghana. *Journal of Plant Pathology & Microbiology.*, **5**(1):1.

Amoako-Attah, I., Akrofi, A. Y., Hakeem, R. B., Asamoah, M., & Kumi-Asare, E. (2016). White thread blight disease caused by *Marasmiellus scandens* (Massee) Dennis Reid on cocoa and its control in Ghana. *African Journal of Agricultural Research.*, 11(50):5064-5070.

Bailey, B. A. and Meinhardt. W. (2016). Cacao Diseases: A History of Old Enemies and New Encounters. Springer international publishing, Switzerland. Pp 450.

Bateman, R. P., Hidalgo, E., García, J., Arroyo, C., Ten Hoopen, G. M., Adonijah, V., & Krauss, U. (2005). Application of chemical and biological agents for the management of frosty pod rot (*Moniliophthora roreri*) in Costa Rican cocoa (*Theobroma cacao*). *Annals of Applied Biology.* 147(2):129-138.

Bowers, J. H., Bailey, B. A., Hebbar, P. K., Sanogo, S., & Lumsden, R. D. (2001). The impact of plant diseases on world chocolate production. *Plant Health Progress.* 2(1): 12.

Cheesman, E. E. (1944). Notes on the nomenclature, classification and possible relationships of cacao populations. *Tropical Agriculture*, 21(8).

Costa, J. C. B., Bezerra, J. L., & Cazorla, I. M. (1996). Controle biológico da vassoura-de-bruxa do cacueiro na Bahia com Trichoderma polysporum. *Fitopatologia Brasileira.*, 21(Suppl.): 397.

Delgado-Ospina, J., Molina-Hernández, J. B., Chaves-López, C., Romanazzi, G., & Paparella, A. (2021). The Role of Fungi in the Cocoa Production Chain and the Challenge of Climate Change. *Journal of Fungi.* 7(3): 202.

Engelbrecht, C. J., Harrington, T. C., & Alfenas, A. (2007). Ceratocystis wilt of cacao—a disease of increasing importance. *Phytopathology.* 97(12), 1648-1649.

Gregory, P. H., & Maddison, A. C. (1981). Epidemiology of Phytophthora on cocoa in Nigeria. *Phytopathological paper.* 25: 23-26.

Guest, D. (2007). Black pod: diverse pathogens with a global impact on cocoa yield. *Phytopathology.* 97(12): 1650-1653.

Guest, D. I., & Keane, P. J. (2018). Cocoa diseases: Vascular streak dieback. Achieving Sustainable Cultivation of Cocoa. Burleigh Dodds Science Publishing, Cambridge.pp 15.

Guest, D., & Keane, P. (2007). Vascular-streak dieback: A new encounter disease of cacao in Papua New Guinea and Southeast Asia caused by the obligate basidiomycete *Oncobasidium theobromae*. *Phytopathology.* 97(12), 1654-1657.

Holderness, M. (1990). Control of vascular-streak dieback of cocoa with triazole the problem of phytotoxicity. fungicides and *Plant Pathology.*, 39(2): 286-293.

Mohanan, R. C., Kaveriappa, K. M., & Nambiar, K. K. N. (1989). Epidemiological studies of *Colletotrichum gloeosporioides* disease of cocoa. *Annals of Applied Biology.* 114(1): 15-22.

Sriwati, R., Chamzurn, T., Soesanto, L., & Munazhirah, M. (2019). Field application of Trichoderma suspension to control cacao pod rot (*Phytophthora palmivora*). *AGRIVITA, Journal of Agricultural Science.* 41(1): 175-182.

Vanegtern, B., Rogers, M., & Nelson, S. (2015). Black pod rot of cacao caused by *Phytophthora palmivora*. *Plant Disease.* 108:1-5.

Chapter - 4

Diseases of Black Pepper *(Piper nigrum* L.) and their Integrated Management

Pranab Dutta, Alinaj Yasin, Arti Kumari,
Madhusmita Mahanta and Anwesha Sharma

College of Post Graduate Studies in Agricultural Sciences
Central Agricultural University, (Imphal), Umiam, Meghalaya

Introduction

Black Pepper, known as the King of Spices, is the most important and most widely used spice in the world. The black pepper of commerce is the dried, mature fruits (commonly called berries) of the tropical, perennial climbing plant *Piper nigrum* L., which belongs to the family *Piper*aceae., Black pepper is a woody climber, rightly called the King of Spices, and its position is supreme among spices. This spice with its characteristic pungency and flavour is an ingredient in many food preparations, and at the dining table, it is the only spice invariably served. The largest producers of pepper in the world are Vietnam and Indonesia followed by India in third (Kueh, *et al.*, 1993). Black pepper is mostly grown in the South Western region of India, comprising of the states of Kerala, parts of Karnataka, Tamil Nadu and Goa, the entire region once known as Malabar (Barber, 1905). Currently, pepper is grown in about 26 countries globally. Production and export of the pepper in the world were dominated by India but the status is losing and there has been a stagnant in the production of pepper at around 50,000 tonnes in the last few years. The major portion of black pepper production in the country is accounted for by Kerala and Karnataka. In the country, Kerala accounts for 75 per cent of the total production (Sadanandan *et al.*, 1992). The yield of pepper is highly affected by the frequent attack of diseases, lack of pest control measures, lack of fertilizer, water facilities, climatic conditions and nature of the soil (Thomas and Menon, 1939; Nambiar *et al.*, 1978;).

Economic Importance of Diseases

Black pepper is affected by several diseases caused by fungi, bacteria, viruses, nematodes and mycoplasma, besides nutritional disorders (Randombage and Bandara, 1984; Nair and Sarma, 1988; Sarma *et al.*, 1994; Ravindran, 2000; Zakaria and Noor, 2020). Crop losses due to diseases and pests are identified as major causes of the low productivity of pepper in India (Menon, 1949; Nambiar, 1978; Sarma *et al.*, 1988a, 1988b; Sarma *et al.*, 1991; Sarma *et al.*, 1992a, 1992c). Crop losses caused by diseases are a major production constraint in all pepper-producing countries. In India, Indonesia and Malaysia, *Phytophthora* foot rot is the major disease. Other diseases include slow decline, anthracnose, viral diseases which are referred to as a stunted disease, stunted growth and wrinkled leaf disease (Sitepu and Kasim, 1991; Kueh and Sim 1992b). In India, although wilt disease was the major disease-causing death of plants, *Phytophthora* as the causal organism was reported only in 1966 by Samraj and Jose. Several diseases were recorded subsequently and later 17 diseases are known to affect pepper. The diseases of pepper are reviewed. Among these diseases, *Phytophthora* foot rot, a slow decline which was previously referred to as "quick wilt" and "slow wilt" respectively, anthracnose and stunted disease cause severe crop losses (Sarma *et al.*, 1992b). The nutrient imbalance between the soil and plant often predisposes the pepper plants to diseases including spike shedding and stem wilting and leaves yellowing. Spike shedding and stem wilting of pepper plants were reported to cause about 30% yield loss of the crop (CPCRI, 1986; IISR1995; 1997; NRCS, 1994).

Major Diseases

1. Foot rot or quick wilt disease of pepper

Economic importance

In India, pepper is grown in the Western Ghats, in the states of Kerala, Karnataka and Tamilnadu. Foot rot has been reported from the introduced areas such as Tripura. The crop losses due to foot rot of pepper are reported to range up to 30 per cent Foot rot takes a heavy toll in all pepper-producing countries and 5–10 per cent loss has been reported in Malaysia and up to 95 per cent loss in individual gardens, a similar situation prevails in India also (Nair *et al.*, 1988). The expressions of the symptom depend upon the site of infection and the extent of the damage. Broadly, infections caused by *Phytophthora* in pepper are classified as aerial and soil infections (Abraham et *al.,* 1996; Sarma and Nambiar, 1982; Sarma *et al.,* 1982). The aerial occurs on the runner shoots, foliage, spikes and branches causing blight, defoliation, spike shedding and die back and many times lead to the death of plants. Infections on the runner shoots often reach the collar causing foot rot (Brahma *et al.,* 1980; Sarma and Anandaraj, 1997).

Symptoms

In foliar infection, one or more dark spots having characteristic fimbriate advancing margins occur in the leaves which later coalesce leading to defoliation even before the lesions spread to the entire lamina. Infections occur both on the tender leaves and shoots and on runner shoots (stolons) that arise at the base of the vines and trail on the ground. The sporulated fungus abundantly forms a white covering on the blighted tender shoot. Infection on tender shoots upon reaching the stem, causes, collar infection and sudden wilting of the plant. Infected spikes cause the blackening of developing fruits and peduncles and eventually spikes are shed. Infection on branches causes drying and defoliation. Stem infection results in yellowing and blighting of stem and complete death of plants. The infected vines may recover after the rains and survive for more than two seasons till the root infection culminates in collar rot and death of the vine .

Infection on below-ground parts such as roots and collars is fatal. Infections of feeder roots cause yellowing, defoliation and drying up of plants due to their rotting and degeneration (De Waard, 1979). Feeder root infection reaches the collar through the main roots and causes foot rot. Pepper being vegetatively propagated, roots arise at each node of the main stem which remains below the soil. If the infection reaches the collar, either through the runner shoots or through the roots closer to the soil level, plants show only wilting symptoms, whereas, infection reaching the collar through the roots of the lower nodes leads to yellowing and defoliation before succumbing to the infection. Such plants may remain alive for 2–3 years (Anandaraj *et al.,* 1996a)

Causal organism

Phytophthora palmivora var. *Piperis* survives in the soil as well as in plant debris (Brasier, 1969). The fungus attacks the root, later spreading to all other parts of the plants. The fungal hyphae are found both intracellular and intercellular in the host and are 2 to 5 μ in thickness. It produces characteristic hyaline, thin-walled sporangia which measure 30-40 X 15-20 μ. (Anandaraj and Sarma, 1990; Kueh and Sim, 1992a).

Disease cycle

Contaminated soil is the main source of inoculum for infection. The inoculum survives in the soil for up to 19 months in the absence of a host plant. The fungus survives as a saprophyte during adverse climatic and soil conditions by means of oospores and chlamydospores. Secondary spread is by sporangia and zoospores, disseminated by irrigation water and rains accompanied by wind.

Epidemiology

Aerial infection results in defoliation and in severe cases death of plants. The spread is dependent on the prevalence of favourable weather. A combination of factors such as daily rainfall of 15.8–23.0 mm, temperature range of 22.7–29.6°C, relative humidity of 81–99 per cent and sunshine of 2.8–3.5 h /day favour the spread of aerial infection. Being perennial, the growth also depends upon the availability of soil moisture. The weather conditions during monsoon along with high soil moisture (>25%), temperatures ranging from 22–29°C, and relative humidity >80 per cent, are favourable for rapid multiplication of the fungus (Kasim and Prayitno, 1979; Nambiar and Sarma, 1982).

Disease management

Integrated disease management involving cultural, chemical and biological control is recommended (Anandaraj and Sarma, 1995; Sarma *et al.*, 1988).

Cultural control

Provision of drainage: Population build-up of *P. palmivora* var. *Piperis* is dependent on weather and is positively correlated with soil moisture. High precipitation during the rainy season leads to waterlogged conditions. Such conditions predispose plants to *Phytophthora* infections.

Phytosanitation: Pepper *Phytophthora* is reported to survive up to 19 months in plant debris

Removal of infected plants would reduce the inoculum level and spread of the fungus.

Shade regulation: During the monsoon season the canopies generate a microclimate under their canopies with high humidity and low temperature, which is ideal for *P. palmivora* var. *Piperis* to multiply and infect. Lopping off the branches during the rainy season is essential to facilitate penetration of sunlight and reduction of high humidity thereby altering the microclimate. The lopped branches could be used as mulch to prevent soil splashes.

Removing weeds and turning the topsoil would help to conserve soil moisture, as the capillary pores are broken and soil moisture removal by weeds during summer is also reduced. Once weeds are removed the saprophytic survival of *P. capsici* on weeds would also be reduced.

Biological control

Soil-borne pathogens are amenable to biological control. In the rhizosphere of pepper several antagonistic microorganisms belonging to *Trichoderma* and *Gliocladium* occur (Anandaraj and Peter, 1996; Cook and Baker, 1983; Graham, 1988; Sarma *et al.*, 1996).

Use of beneficial microorganisms

Several beneficial organisms like vesicular-arbuscular mycorrhizae (VAM) are associated with pepper (Graham, 1982). VAM inoculation has been found to enhance rooting, and growth and suppress root damage caused by *P. palmivora* var. *Piperis*, *P. capsici*, *Radopholus similis* and *Meloidogyne incognita* under artificial inoculation and under field conditions (Sivaprasad *et al.*, 1990a, 1995; Anandaraj and Sarma, 1994; Anandaraj *et al.*, 1996b; Kueh and Sim, 1992c).

Organic amendments

Organic matter such as neem oil cake soybean meal, ground nut cake, coconut cake and chicken manure are added to the soil to supplement nutrition and enhance the growth of saprophytes (Anandaraj and Leela, 1996; Nair *et al.*, 1993).

Chemical control

Although the disease occurs every year during the rainy season, a fixed fungicide scheduling is advocated against both aerial infection and collar infection. The control measures include the combination of Metalaxyl M 4% and Mancozeb 64% WP @ 0.25%, 2 or 3 1 per vine. Another combination is Metalaxyl 8% + Mancozeb 64% WP @ 0.125 %, 2 or 5 1 per vine is effective against the pathogen. It is difficult to control the root rot phase of the disease, but drenching with a 1 per cent Bordeaux mixture is useful in protecting the plants to some extent (Anandaraj and Sarma, 1991; Kueh *et al.*, 1992; Nair and Sasikumaran, 1991).

2. *Fusarium* Wilt of Pepper

Economic importance

In Brazil, the major disease that is creating havoc in pepper cultivation is the *Fusarium* wilt, and this disease is causing severe losses in that country. *Fusarium* infection in the plantation is reported to reduce the economic life of the plantation from 20 to 6–8 years and the productivity per plant from 3.0–1.5 kg (Nambiar and Sarma, 1977).

Symptoms of root rot and stem blight

The symptoms of *Fusarium* infection are root rot leading to flaccidity and yellowing of foliage. At the early age of pathogen establishment, the black pepper produces a normal appearance but after 4 months of infection, the yellowing symptom will start to appear on the vines and spread to the top (Duarte and Albuquerque, 1991). As the infection reach an advanced stage, the plant suffers from defoliation. Reddish discolouration on the xylem vessel could be observed when the stem is cut. This disease can occur in both young and old vines. At a certain stage, main roots lose their feeder roots causing dieback and death of vines (Balakrishnan *et al.*, 1986, Beckman, 1987, Huisman, 1982).

Causal organism

The organism is identified as *Fusarium solani* f. sp. *Piperis*. The perfect stage has been identified as *Nectria haematococca* f. sp. *Piperis*. In affected plants, the fungus produces abundant perithecia (Shahnazi *et al.*, 2012).

Disease cycle

Fusarium species can be soil-borne, airborne, carried in plant residue and can be recovered from any part of the plant from the end of deepest roots up until the top of the flower. *Fusarium* spp. usually, block xylem vessels result to plant wilt and die. *Fusarium* produces macroconidia and microconidia including mycelia and chlamydospores which are responsible for infecting the host plant. The life cycle of this pathogen can be classified as dormant, parasitic and saprophytic stages. The dormant stage involves inhibition and germination of the resting phase in soil. While parasitic stage includes penetration and colonization at the vascular system of plant roots including movement in the xylem vessel resulting in the death of the host plant. The saprophytic stage is a resting structure in the dead host. Mycelia, chlamydospores, macroconidia and microconidia are available in the infected soil during the dormant phase due to mycisstasis. Later the parasitic stage happens once any of the propagules penetrates the host through the formation of a crack or wound at the root parts (Butler, 1906; Nambiar and Sarma, 1979).

Disease management

Cultural control

Fusarium infection in plantations is reported to reduce the economic life of plantations from 20 to 6–8 years and the productivity per plant from 3.0–1.5 kg.

Phytosanitary

Measures like removing infected plants and burning and application of 200 g Calcium cyanamide per mound are practised.

Chemical control

Measures with both systemic and contact fungicides are resorted to depending on the nature of the damage. Benomyl, carbendazim and thiobendazole at 0.5 and 0.6 per cent are reported to be effective against *Fusarium*. Prophylactic application of fungicides at the fortnightly interval, systemic fungicide followed by contact fungicide to the foliage and soil drenching with Benomyl (0.05%) and Thiobendazole (0.06%) were found effective in controlling the disease.

3. Anthracnose of Pepper

Economic importance

Anthracnose in pepper is referred to as "pollu" disease in India, which means hollow fruits and as blackberry disease in Malaysia and Indonesia. This occurs both on the leaves and on spikes. Although sporadic in nature in the major pepper-growing state of Kerala, this disease is becoming severe in parts of Karnataka where pepper is grown on shade trees in coffee plantations. The severity of the disease varies and a range of 28–34 per cent has been reported causing a crop loss of 1.9–9.5 per cent.

Symptoms

Colletotrichum symptoms on black pepper can be figured out once all of the whole berries at the spike turn black and fall. The symptoms of fruits depend on the stage of maturity. On younger fruits, infection leads to blackening. On mature fruits, brownish lesions are formed. Necrosis started at the end tips of berry stalks soon dark lesions appeared on the other infected berries and the centre of each berry started to crack and fall. If the infection occurs on the stalk end of the spike, the entire spike is shed prematurely, and brown lesions are produced on the leaves (Sarma and Anandaraj, 1992; Wilson, 1960).

Causal organism

The fungus *Colletotrichum necator* is the cause of this disease. From India, *Colletotrichum* sp., *C. gloeosporioides* are also recorded. The fungus is reported to survive on *Dioscorea triphylla* as an alternate host (Barimani *et al.,* 2013).

Disease cycle

Colletotrichum infects a wide variety of plant crops causing anthracnose disease and having both sexual and asexual phases. Primary infection was formed on the epidermal cell and does not kill the plant but the necrotrophic stage of secondary infection will invade and kill the plant. Quiescent lifestyle refers to the pathogen that exists dormant in the first phase before switching to the active phase. *Colletotrichum* often remains dormant until the symptom appears during the harvest, storage, transportation and sales of the product. An endophytic lifestyle where fungi live within plant cells as symbionts without causing diseases. Usually, *Colletotrichum* species exist as endophytes for most of their life cycle in every type of crop.

Epidemiology

The infection of spores starts with rain splashing onto pepper plants and dispersing to other berries and in some other cases spores being dispersed through unsanitized equipment during handling of the infected plant. *Colletotrichum* can survive in wet and warm conditions with temperatures of 27 °C and 80% humidity.

Disease management

Chemical control

As the disease occurs during the rainy season, Bordeaux mixture (1%) spray is recommended. Apart from Bordeaux mixture spray, Difolatan at 0.2 per cent was also reported to give adequate control. Benomyl, Thiophanate ethyl, Thiophanate methyl, Carbendazim, Captofol and Tridemefron were also effective in controlling this disease. Anthracnose is reported to be reduced when 40 per cent shade is provided by using shade nets (Nair *et al.,* 1987).

4. Stunted Disease

Economic importance

This disease, also known as little leaf disease, was recorded in 1975 in the Idukki district of Kerala which is one of the major pepper-producing areas. Earlier it was sporadic but is now found in all major pepper-growing tracts of Kerala and Karnataka. A survey has indicated that the number of plants affected in the Wynad district of Kerala ranged from 0.6 to 18.6 per cent. This disease is known as mosaic disease and as "little leaf" in Sri Lanka, mosaic disease in India, "wrinkled leaf disease" in Malaysia and as "stunted disease" in Indonesia (Lockhart *et al.*, 1997; Prakasam *et al.*, 1990).

Symptoms

The leaves become narrow and leathery in texture, with puckering and chlorotic streaks, internodes become shortened and the branches appear as witch's broom, at times the leaves show narrowing and vein banding. The appearance of symptoms is pronounced in neglected gardens and older plants recovering after a low decline infection. In the nurseries, chlorotic streak and vein banding are more common. The four categories of symptoms namely, stunting, reduction in the size of internodes, narrowing of leaves, marginal necrosis and chlorosis (Eng *et al.*, 1993).

Causal organism

Transmission tests carried out in India have shown that the disease is graft transmissible. ELISA tests indicated that the disease is caused by the cucumber mosaic virus (CMV). Studies with immunosorbent electron microscopy, and transmission electron microscopy using ultra-thin sections also did not reveal the presence of virus in infected tissues. However, purified extracts have been reported to contain bacilliform virus particles measuring 120×30 nm, isometric virus of 30 nm and rod-shaped virus-related proteins. The virus is transmissible mechanically as well as by citrus mealy bug.

Disease management

Cultural control: Presently, the use of virus-free healthy planting material following regular inspection and removal of infected plants; the removed plants may be burnt or buried deep in soil is recommended. Insects such as aphids and mealy bugs on the plant or standards should be controlled with insecticide spray.

Minor Diseases

5. Phyllody

This disease has been reported in from Wynad district of Kerala which is a major pepper growing area, this is also called antholysis.

Symptoms

The spikes and flowers are converted into leaf-like structures instead of normal flowers and the stalk of the affected spikes is also elongated. The leaves of the affected plants become smaller and show chlorosis. In the same vine, both normal spikes and berries are also produced (Sarma *et al.*, 1988; Sarma *et al.*, 1988, Sarma and Anandaraj, 1992).

Causal organism

Based on electron microscopic studies, the disease is reported to be caused by phytoplasma.

Disease management

As the disease is confirmed to be caused by phytoplasma, removal of affected plants and following quarantine measures are suggested.

6. Bacterial Leaf Spot

Symptom

The symptoms appear as small water-soaked lesions on the leaf lamina and margins which later become dark with a chlorotic halo and times cause defoliation (Mathew *et al.*, 1978, 1979).

Causal organism

This disease is caused by *Xanthomonas compestris* pv. *betlicola* and occurs sporadically in certain pockets and in the vicinity of other *Piper* spp. The causal organism is *Xanthomonas compestris* (Bhale *et al.*, 1984).

Management

Chloramphenicol at 200 ppm was reported to be effective in inhibiting the growth of the bacterium. However, control measures to prevent this disease in the field are not available.

7. Thread Blight

This disease occurs on pepper leaves and spikes. The fungus grows underneath the leaves and the on stem causing drying up of leaves and spikes.

Pellicularia filamentosa (*Corticium solani*) is reported as the causal organism.

8. Stump Rot

The occurrence of stump rot caused by *Rosellina bunodes* was recorded in Wynad areas of Kerala. The fungus affects the root system which results in the drying up of plants. The fungus also affects *Grevillea robusta* on which pepper plants are trailed.

Isolation of affected plants by making trenches is recommended to prevent the spread of the disease to adjacent plants.

9. Red Rust

Red rust caused by the alga *Cephaleuras mycoidea* occur on older leaves in certain pockets. This was reported to cause black fruit in Malaysia. This occurs on the surface of leaves, thereby cutting off light required for photosynthesis. The exact nature of the damage has not been assessed. When the growth occcurs on the spikes and fruits the appearance of the berries is affected and the quality is reduced.

10. Velvet Blight

The velvet blight is caused by *Septobasidium* sp. The fungus grows on the surface of the fruits and forms a coating but does not penetrate the fruits. Whereas, when it occurs on the branches it results in dieback symptoms. Pruning the affected branches would prevent the spread of the fungus.

Future Approaches

Accurate identification of causal pathogens along with knowledge of handling the impacted diseases are essential in developing specific management strategies to control each disease. The application of organic manures along with biocontrol agents would prevent the population build-up of *P. capsici* in soil and protect the roots against damage. Present research efforts are concentrated on host resistance, by understanding the mechanism of resistance and incorporating the same through biotechnological means, so that host resistance becomes an important component of the integrated management of diseases of pepper.

Conclusion

India has a wide variety of crops for commodities and some of the growers prefer to practice mixed cropping in one piece of land in maximizing the potential use of land and their profit. These situations have to lead to a high impact of diseases. In pepper, all the diseases described above occur during the wet monsoon season under Indian conditions. In the management of diseases, the unfixed fungicide schedule is followed when favourable weather conditions are attained during the monsoon period every year leading to occurring of diseases. As a result, the use of high doses and unrecognized chemicals to control black pepper diseases has created resistance to fungicides.

References

Abraham, J., Anandaraj, M., Ramana, K.V., and Sarma, Y.R. (1996). A simple method for indexing *Phytophthora* and nematode infections in black pepper (*Piper nigrum* L). *J. Spices Aromatic Crops*, 5: 68–71.

Anandaraj, M. and Sarma, Y.R. (1990). A simple baiting technique to detect and isolate *Phytophthora capsici* (*P. palmivora* MF4) from the soil. *Mycol. Res.*, 94: 1003–1004.

Anandaraj, M., and Sarma, Y.R. (1991). Use of baits for assaying chemicals applied as soil drench to control *Phytophthora* foot rot of black pepper (*Piper nigrum* L.). *Indian Phytopath.*, 44, 543–544.

Anandaraj, M., and Sarma, Y.R. (1994). Effect of Vesicular arbuscular mycorrhiza on rooting of black pepper (*Piper nigrum* L.). *J. Spices Aromatic Crops*, 3: 39–42.

Anandaraj, M., and Sarma, Y.R. (1995). Diseases of black pepper (*Piper nigrum* L.) and their management. *J. Spices Aromatic Crops*, 4: 17–23.

Anandaraj, M., and Leela, N.K. (1996). Toxic effect of some plant extracts on *Phytophthora capsici*, the foot rot pathogen of black pepper. *Indian Phytopath.*, 49: 181–184.

Anandaraj, M., and Peter, K.V. (1996). Biological Control in Spices. Indian Institute of Spices Research, Calicut, India, pp. 52.

Anandaraj, M., Ramana, K.V., and Sarma, Y.R. (1996a). Role of *Phytophthora capsici* in the slow decline disease of black pepper. *J. Plantation Crops*, 24: 166–170.

Anandaraj, M., Ramana, K.V., and Sarma, Y.R. (1996b). Sequential inoculation of *Phytophthora capsici*, *Radopholus similis* and *Meloidogyne incognita* in causing slow decline of black pepper. *Indian Phytopath.*, 49: 297–299.

Balakrishnan, R., Anandaraj, M., Nambiar, K.K.N., Sarma, Y.R., Brahma, R.N., and George, M.V. (1986). Estimates on the extent of loss due to quick wilt disease of black pepper (*Piper nigrum* L.) in Calicut district of Kerala. *J. Plantation Crops*, 14: 15–18.

Barber, C.A. (1905). The Government pepper farm in Malabar. *Tropic. Agric.*, 25: 564.

Barimani, M., Pethybridge, S.J., Vaghefifi, N., Hay, F.S., and Taylor, P.W.J. (2013). A new anthracnose disease of pyrethrum caused by *Colletotrichum tanaceti* sp. nov. *Plant Pathol.* 62, 1248e1257.

Beckman, C. H. (1987). The Nature of Wilt Diseases of Plant. St. Paul, Minnesota: *American Phytopathol. Soc. Press*, p. 175.

Bhale, N.B., Nayak, M.I., and Chourasia, R.D. (1984). New bacterial disease of betel vine. *Indian Phytopath.*, 37: 373.

Brahma, R.N., Nambiar, K.K.N., and Sarma, Y.R. (1980). Basal wilt of black pepper and its control. *J. Plantation Crops*, 8: 107–109.

Brasier, C.M. (1969). Formation of oospores in vivo by *Phytophthora palmivora*. *Trans. Br. Mycol. Soc.*, 52: 273–279.

Butler, E.J. (1906.) The wilt disease of Pigeon pea and Pepper. *Agric. J. India*, 1: 25–26.

Central Plantation Crops Research Institute (1986). Annual Report for 1985, Central Plantation Crops Research Institute, Kasaragod, pp. 17.

Cook, R.J., and Baker, R. (1983). The Nature and Practice of Biological Control of Plant Pathogens, *American Phytopath. Soc.*, St. Paul, Minnesota, pp. 53.

De Waard, P.W.F. (1979). Yellow disease complex in black pepper on the island of Bangka, Indonesia. *J. Plantation Crops*, 7: 42–49.

Duarte, M.L.R., and Albuquerque, F.C. (1991). *Fusarium* disease of black pepper in Brazil. In: Y.R.Sarma and T.Premkumar (eds.), Black Pepper Diseases, National Research Centre for Spices, Calicut, India., pp. 39–54.

Eng, L., Jones, P., Lockhart, B., and Martin, R.R. (1993). Preliminary studies on the virus diseases of black pepper in Sarawak. In: Ibrahim, M.Y., ,.F.J., and Ipor. I.P. (eds.), The Pepper Industry: Problems and Prospects, Universiti Pertanian Malaysia, Bintulu campus, Malayasia, pp. 44–46.

Graham, J.H. (1982). Effect of citrus root exudates on germination of chlamydospores of the vesicular-arbuscular mycorrhizal fungus *Glomus epigaeum*. *Mycologia*, 74: 831–835.

Graham, J.H. (1988). Interactions of mycorrhizal fungi with soil borne plant pathogens and other organisms, An introduction. *Phytopathology*, 78: 365–366.

Graham, J.H., and Egel, D.S. (1988). *Phytophthora* root rot development on mycorrhizal and phosphorus-fertilized non-mycorrhizal sweet orange seedlings. *Plant Dis.*, 72: 611–614.

Holliday, P. (1998). A Dictionary of Plant Pathology (2nd ed.)

Huisman, O. C. (1982). Interactions of root growth dynamics to epidemiology of root invading fungi. Annual Review of Phytopathol., 20: 303-327.

Indian Institute of Spices Research (1995). Annual Report for 1994–95, Calicut, India. pp. 89.

Indian Institute of Spices Research (1997). Research Highlights for 1996–97, Calicut, India, pp. 10.

Kasim, R., and Prayitno, S. (1979). Factors influencing growth and sporulation of *Phytophthora capsici* from *Piper nigrum* L. *Pemeritan Lembaga Penelitian Tanaman Industri*, Indonesia, 34: 41.

Kueh T.K., Lim, J.L., and Nordmeyer, D. (1992). Soil application of Ridomil R 5G for the control of *Phytophthora* foot rot in black pepper. In: Ibrahim, M.Y., Bong C.F.J., and Ipor, I.P. (eds.), *Proc. Pepper Industry: Problems Prospects*, Univeristi Pertanian Malaysia, Sarawak, Malaysia, pp. 169–171.

Kueh, T.K., and Sim, S.L. (1992a). Etiology and control of *Phytophthora* foot rot of black pepper in Sarawak, Malaysia. In: Wahid, P., Sitepu, D., Deciyanto, S., and Superman, U. (eds.), *Proc. International Workshop on Black Pepper Diseases*, Bander Lampung Indonesia, Institute for Spice and Medicinal Crops, Bogor, Indonesia, pp. 155–162.

Kueh, T.K., and Sim, S.L. (1992b). Occurrence and management of wrinkled-leaf disease of black pepper. In: Wahid, P., Sitepu, D., Deciyanto S., and Superman, U. (eds.), *Proc. International Workshop Black Pepper Diseases*, Bander Lampung Indonesia, Institute for Spice and Medicinal Crops, Bogor, Indonesia, pp. 227–233.

Kueh, T.K. and Sim, S.L. (1992c) Slow decline of black pepper caused by root knot nematodes. In: Wahid, P., Sitepu, D., Deciyanto S., and Superman, U. (eds.), *Proc. International Workshop on Black Pepper Diseases*, Bander Lampung Indonesia Institute for Spice and Medicinal Crops, Bogor, Indonesia, pp. 198–207.

Kueh, T.K., Fatimah, O. and Lim, J.L. (1993) Management of black berry disease of black pepper. (eds.), In: Wahid, P., Sitepu, D., Deciyanto S., and Superman, U. (eds.), *The Pepper Industry: Problems and Prospects*, Universiti Pertanian Malaysia, Sarawak, Malaysia, pp. 162–168.

Kueh, T.K., Gumbek, M. Wong, T.H., Chin, S.P. (1993). A field guide to diseases, pests and nutritional disorders of black pepper in Sarawak. Agricultural Research Centre Semongok, Department of Agriculture Kuching. Lee Ming Press, Sarawak.

Lockhart, B.E.L., Kiratiya Angul, K., Jones, P., Eng, L., Silva, P.De., Olszewski, N.E., Lockhart, Deema, N., and Sangalang, J. N. (1997). Identification of *Piper* Yellow Mottle Virus, a mealy bug transmitted Badna virus infecting *Piper* Spp. in South East Asia. *European J. Plant Path.* 103: 303–311.

Manjunath, A., and Bagyaraj, D.J. (1982). Vesicular arbuscular mycorrhizas in three plantation crops and cultivars of field bean. *Curr. Sci.*, 51: 707–709.

Mathew, J., Cherian, M.T., and Abraham, K. (1978). *Piper nigrum* a new host of *Xanthomonas betlicola. Curr. Sci.*, 47: 956–957.

Mathew, J., Abraham, K., and Wilson, K.I. (1979). In vitro effects of certain antibiotics against *Xanthomonas betlicola* causing leaf spot of pepper. In: Venkata, C.S., Ram (ed.), Proc. Second Plantation Crops Symposium (PLACROSYM II), Indian Society for Plantation Crops, Central Plantation Crops Research Institute, Kasaragod, pp. 401–402.

Menon, K.K. (1949). The survey of pollu and root diseases of pepper. *Indian J. Agric. Set.*, 19: 89–132.

Nair, M.K., and Sarma, Y.R. (1988). International pepper community workshop on joint research for the control of black pepper diseases. *J. Plantation Crops*, 16: 146–149.

Nair, P.K.U., Sasikumaran, S., and Pillay, V.S. (1987). Time of application of fungicides for control of anthracnose disease of pepper (fungal pollu). *Agric. Res. J.* Kerala, 25: 136–139.

Nair, P.K.U., Sasikumaran, S., Pillai, V.S., and Prasadarao, G.S.L.H.V. (1988). Influence of weather on the incidence of foot rot disease of black pepper (*Piper nigrum* L.) In: Prasadaro, G.S.L.H.V., and Nair, R.R. (eds.), Agrometerology of Plantation Crops, Kerala Agricultural University, Trichur, pp. 98–103.

Nair, P.K.U., and Sasikumaran, S. (1991). Effect of some fungicides on quick wilt (foot rot) disease of black pepper. *Indian Cocoa, Arecanut Spices J.*, 14: 95–96.

Nair, P.KU., Mammootty, K.P., Sasikumaran, S., and Pillay, V.S. (1993). *Phytophthora* foot rot of black pepper (*Piper nigrum*, L.) a management study with organic amendments. *Indian Cocoa, Arecanut and Spices J.*, 17: 1–2.

Nambiar, K.K.N., and Sarma, Y.R. (1977). Wilt diseases of black pepper. *J. Plantation Crops*, 5: 92–103.

Nambiar, K.K.N., (1978). Diseases of pepper in India. In: Nair, M.K., and Haridasan, M. (eds.), *Proc. National Seminar on Pepper*, Calicut, Central Plantation Crops Research Institute, Kasaragod, pp. 11–14.

Nambiar, K.K.N., and Sarma, Y.R. (1979). Factors associated with slow wilt of pepper. In: C.S.Venkata Ram (ed.), Proc. Second Plantation Crops Symposium (PLACROSYM II), *Indian* Society for Plantation Crops, Central Plantation Crops Research Institute, Kasaragod, pp. 386–389.

Nambiar, K.K.N., and Sarma, Y.R. (1982). Some aspects of epidemiology of foot rot of black pepper. In: Nambiar, K.K.N. (eds.), *Phytophthora* Diseases of Tropical Cultivated Plants. Central Plantation Crops Research Institute, Kasragod, pp. 225–231.

Nambiar, P.K.V., Sukumara Pillay, V., Sasikumaran, S., and Chandy, K.C. (1978). Pepper Research at Panniyur-A Resume. *J. Plantation Crops*, 6: 4–11.

National Research Centre for Spices (NRCS) (1994). Annual Report for 1993–94, Calicut, Kerala, pp. 7.

Okungbowa, F. I., and Shittu, H. O. (2012). *Fusarium* wilts: An overview. *Environ. Res. J*, 6(2): 83-102.

Prakasam,V., Subbarajan, K.T., and Bhaktavatsalu, C.M. (1990). Mosaic disease—a new record in black pepper in lower Pulneys. *Indian Cocoa, Arecanut and Spices J.*, 13: 104.

Randombage, S., and Bandara, J.M.R.S. (1984). Little leaf disease of *Piper nigrum* in Sri Lanka. *Plant Pathol.*, 33: 479–482.

Ravindran, P.N. (2000). Diseases of black pepper. In: Anandaraj, M. Black pepper. Overseas Publishers Association, Amsterdam, Netherlands.

Sadanandan, A.K., Abraham, J., and Anandaraj, M. (1992). Impact of high production technology on productivity and disease of black pepper (*Piper nigrum* L.). *J. Plantation Crops.*, 20 (Suppl), 384–387.

Samraj, J. and Jose, P.C. (1966). A *Phytophthora* wilt of Pepper (*Piper nigrum*). *Sci Cult.*, 32: 90–92.

Sarma,Y.R., Ramachandran, N., and Nambiar, K.K.N. (1982). Morphology of black pepper *Phytophthora* isolates from India. In: Nambiar, K.K.N. (ed.), *Phytophthora* Diseases of Tropical Cultivated Plants, Central Plantation Crops Research Institute, Kasaragod, pp. 233–235.

Sarma, Y.R., and Nambiar, K.K.N. (1982). Foot rot disease of black pepper (*Piper nigrum* L.) In: Nambiar, K.K.N. (ed.), *Phytophthora* Diseases of Tropical Cultivated Plants, Central Plantation Crops Research Institute, Kasaragod, pp. 209–224.

Sarma, Y.R., Ramachandran, N., and Anandaraj, M. (1988). Integrated disease management of 'quick wilt' (foot rot) of black pepper caused by *Phytophthora palmivora* MF4. *Coffee Res. (Suppl)*, 18: 68–72.

Sarma,Y.R., Solomon, J.J., Ramachandran, N., and Anandaraj, M. (1988). Phyllody disease of black pepper (*Piper nigrum* L.). *J. Plantation Crops*, 16: 69–72.

Sarma,Y.R., Premkumar,T., Ramana, K.V., Ramachandran, N., and Anandaraj, M. (1988a). Disease and pest management in black pepper nurseries. *Indian Cocoa, Arecanut Spices J.*, 11: 45–49.

Sarma,Y.R., Ramachandran, N., Anandaraj, M., and Ramana, K.V. (1988b). Disease management in black pepper. Indian Cocoa, Arecanut and Spices J., 11: 123–127.

Sarma, Y.R., Ramachandran, N., and Anandaraj, M. (1991). Black pepper diseases in India. In: Sarma, Y.R. and Premkumar, T. (eds.), Diseases of Black Pepper, National Research Centre for Spices, Calicut, pp. 55–101.

Sarma, Y.R. and Anandaraj, M. (1992). Anthracnose and phyllody diseases of black pepper in India. In: Wahid, P., Sitepu, D., Deciyanto, S., and Superman, U. (eds.), Proc. The International Workshop on Black Pepper Diseases, Bander Lampung, Indonesia, Institute for Spice and Medicinal Crops, Bogor, Indonesia, pp. 211–214.

Sarma, Y.R., Anandaraj, M., and Devasahayam, S. (1992a). Diseases of unknown etiology of black pepper (Piper nigrum L.). In: Wahid, P., Sitepu, D., Deciyanto, S., and Superman, U. (eds.), Proc. The International Workshop on Black Pepper Diseases, Bander Lampung Indonesia. Institute for Spice and Medicinal Crops, Bogor, Indonesia, pp. 215–217.

Sarma, Y.R., Anandaraj, M., and Ramachandran, N. (1992b). Recent advances in Phytophthora foot rot research in India and the need for a holistic approach. In: Wahid, P., Sitepu, D., Deciyanto, S., and Superman, U. (eds.), Proc. The International Workshop on Black Pepper Diseases, Bander Lampung, Indonesia, Institute for Spice and Medicinal Crops, Bogor, Indonesia, pp. 133–143.

Sarma, Y.R., Anandaraj, M., and Ramana, K.V. (1992c). Present status of black pepper diseases in India and their management. In: Wahid, P., Sitepu, D., Deciyanto, S., and Superman, U. (eds.), Proc. The International Workshop on Black Pepper Diseases, Bander Lampung, Indonesia, Institute for Spice and Medicinal Crops, Bogor, Indonesia, pp. 67–78.

Sarma, Y.R., Anandaraj, M., and Venugopal, M.N. (1994). Diseases of spice crops. In: Chadha, K.L., and Rethinam, P. (eds.), Advances In Horticulture 10(2), Malhotra Publishing House, New Delhi, pp. 1015–1057.

Sarma, Y.R., Anandaraj, M., and Venugopal, M.N. (1996). Biological control of diseases in spices. In: Anandaraj, M. and Peter, K.V. (eds.), Biological Control in Spices, Indian Institute of Spices Research, Calicut, pp. 1–19.

Sarma, Y.R., and Anandaraj, M. (1997). Phytophthora foot rot of black pepper. In: Agnihotri, V.P., Sarbhoy, A.K., and Singh, D.V. (eds.), Management of Threatening Diseases of National Importance, Malhotra Publishing House, India, pp. 237–248.

Shahnazi. S, Meon. S, Vadamalai, G., Ahmad, K., and Nejat, N. (2012). Morphological and molecular characterization of *Fusarium* spp. associated with yellowing disease of black pepper (*Piper nigrum* L.) in Malaysia. *J. Gen. Plant Pathol.*, 78: 160. https://doi.org/10. 1007/s10327-012- 0379-5.

Sitepu, D., and Kasim, R. (1991). Black pepper diseases in Indonesia and their control strategy. In Sarma Y.R., and Premkumar, T. (eds.), Black Pepper Diseases, National Research Centre for Spices, Calicut, pp. 13–28.

Sivaprasad P., Jacob A., Nair, S.K., and George, B. (1990a). Influence of VA mycorrhizal colonization on root knot nematode infestation in *Piper nigrum* L. In: Jalali. B.L., and Chand, H. Current Trends in Mycorrhizal Research. Proc. National Conference on Mycorrhiza, 14– 16 Feb 1990, Haryana Agricultural University, Hissar, pp. 110–111.

Sivaprasad, P., Robert, C.P., Vijayan, M., and Joseph, P.J. (1995). Vesicular-arbuscular mycorrhizal colonization in relation to foot rot disease intensity in black pepper. In Adholeya, A., and Singh, A. (eds.), Mycorrhizae-Biofertilizers For The Future, Tata Energy Research Institute, New Delhi, pp. 137–140.

Thomas, K.M., and Menon, K.K. (1939). The present position of pollu disease of pepper in Malabar. *Madras Agric. J.*, 27: 347–356.

Wilson, K.I. (1960). A new host of *Colletotrichum necator* from Kerala. *Sci. Cult.*, 25: 604–605.

Zakaria, S.N., and Noor., N.B. (2020). A Review On Major Fungus Associated With Black Pepper (*Piper nigrum* L.), Diseases In Malaysia. *International J. Scientific & Engineering Research*.

Chapter - 5

Diseases of Turmeric (*Curcuma longa* L) and their Integrated Management

Madhusmita Mahanta, Pranab Dutta, Alinaj Yasin, Anwesha Sharma and Arti Kumari

School of Crop Protection, College of Post Graduate Studies in Agricultural Sciences Central Agricultural University (Imphal), Umiam - 793103, Meghalaya

Introduction

Turmeric; *Curcuma longa* L., is an annual rhizomatous spice crop native to SE Asia. The turmeric rhizomes are used as a natural antiseptic and have anti-inflammatory, carminative and anti-helminthic properties. Besides, it serves religious, culinary as well as cosmetic purposes. Turmeric is also known as the 'Golden Spice of Life' as well as the 'Indian saffron. India dominates the international turmeric market in production, consumption and export. The crop covers an area of 2.46 lakh hectares with 9.39 lakh tonnes of production. In India, turmeric occupies approximately 6 per cent of the total area under spices and condiments. Further, it contributes 10 per cent share to total spices and condiments production in the country. Turmeric rhizomes contain several pigments among which curcumin is the major one, imparting the medicinal properties as well as the characteristic yellow colour of turmeric (Anonymous, 2020). Indian turmeric has a huge demand in the international spice market because of the high curcumin content in the rhizomes. However, the crop suffers from several diseases in the field, which impacts its yield and thereby affects the economic return (Dohroo, 2007). This chapter encompasses the major diseases of turmeric with their effective integrated disease management practices.

1. Rhizome rot/ soft rot

Symptoms

The disease appears as a soft watery lesion at the collar region of the plant. Drying leaves from the margin is observed as the foliar symptom. The root system is adversely reduced, and rhizome development is less. In the advanced stage of disease development, rotting of rhizomes that emit a foul smell can be observed. In severe conditions, the plant withers and dies (Singh, 2010).

Causal organism: *Pythium aphanidermatum, P. graminicolum*

Disease cycle

The primary infection occurs from the oospores or chlamydospores present in the soil or from the diseased rhizomes used for planting. The pathogen multiplies in the monsoon season. Secondary spread is caused by the conidia.

Epidemiology

High soil moisture and high-temperature 30^0C are the predisposing factors to the disease. The disease spreads faster with the onset of SW monsoon. Young turmeric plants are more susceptible to this disease. Further, nematode infestation and waterlogged fields favour disease development.

Integrated management

Integrated disease management involves the use of healthy and disease-free rhizomes for planting. As the disease is soil-borne, maintaining field sanitation is an important aspect of disease management.

2. Leaf spot

Symptoms

The disease manifests as elliptical to oblong grey spots on leaves with a reddish-brown margin and a yellow halo surrounding the spot. At the center of the spot, acervuli appear in concentric circles as black dots. As the disease progress, the spots elongate and may coalesce to give a blighted appearance. The spots are usually observed to appear on the leaf blade, but in severe conditions, they can be seen on the leaf sheath as well. The leaf spot infected turmeric leaves dry up and give a parched-up appearance to the field (Mahanta *et al.*, 2021).

Causal organism: *Colletotrichum capsici, C. gloeosporioides*

Disease cycle

It is primarily a debris-borne disease. However, the dormant structures of the pathogen that survives in the infected turmeric rhizomes sometimes cause the initial infection. Secondary infection occurs in the conidia that spread to the nearby plants/ field by the rain splashes in humid and rainy weather.

Epidemiology

The pathogen favours 21-30°C with an optimum 25°C temperature for causing infection. The prevalence of rainy and humid weather is necessary for the incidence and dissemination of the leaf spot disease. Leaf wetness helps in the germination of conidia, thereby, assisting in disease development. The disease incidence is maximum during the rainy season in the month of August and September.

Integrated management

Integrated disease management involves the use of healthy and disease-free turmeric rhizomes for planting with recommended spacing. As the disease is mainly debris borne, therefore, the diseased turmeric leaves should be removed and burned to prevent the inoculum build-up in the field. Also, solanaceous crops such as chilli that can act as a reservoir of the pathogen inoculum should not be grown along with the main crop. Further, the use of shade crops (e.g., maize, pigeon pea etc.) is reported to reduce the disease incidence. Rhizome treatment with biocontrol agents such as *Trichoderma harzanium* along with a foliar spray of *T. harzanium* -based bioformulation is effective in reducing the incidence of leaf spot disease. The use of chemical fungicides such as Bevistin @ 0.3% for rhizome treatment and foliar spray @ 0.1% is also recommended in severe cases of disease outbreaks (Dutta *et al.,* 2017).

3. Leaf blotch

Butler in the year 1911 first reported the leaf blotch disease from Gujarat and the Saharanpur region of India.

Symptoms

The disease manifests in the form of numerous small (1-2mm) dirty yellow spots that are mostly confined to the lower leaves of the turmeric plant. The spots appear predominantly on the upper leaf surface. The infected leaves become distorted and the colour appears reddish brown as compared to the healthy leaves. The spots coalesce freely and in severe case, several blacks to brown spot spots appear on both the leaf surfaces of lower leaves. The disease is also responsible for causing yield loss of rhizomes as it hampers the photosynthetic ability

Causal organism: *Taphrina maculans*

Disease cycle

The pathogen survives the summer months on the crop debris as well as on soil as dormant spores which germinates on getting congenial environment. Secondary infection occurs by the ascospores discharged from the infected leaves and causes severe spotting on the fresh and healthy leaves.

Epidemiology

The pathogen favour high humidity (80% RH) and low temperature (21-23°C) for causing infection and disease development. Subsequent spread occurs in cool and humid weather with high soil moisture along with high inoculum potential of the pathogen.

Integrated management

Disease-resistant turmeric cultivar can be used to prevent the occurrence of leaf blotch disease. Field sanitation and crop rotation with the non-host crop are recommended as preventive measures for blotch disease. The use of *Trichoderma harzianum*-based bioformulation for rhizome treatment and as a foliar spray at an interval of 21 days after planting is effective in managing the disease organically. Further, Foliar spray with Mancozeb or COC @ 0.1-0.2% after the appearance of the first symptom at an interval of 15 days is also recommended as a chemical control measure.

Conclusion

Turmeric is an important crop in the country used on several auspicious occasions. The diseases faced by the crop can be prevented or managed effectively if standard cultivation practices are followed. The use of shade trees, field hygiene, and clean intercultural operation play a key role in preventing many diseases. Proper fertilization with biofertilizers and vermicompost enriched with biological control agents like *Trichoderma* is considered an integral part of turmeric disease management. The use of compatible biopesticides and agrochemicals can provide added benefits to the crop.

References

Anonymous (2020). Turmeric outlook report- January to December 2020. Agricultural Market Intelligence Centre, ANGRAU, Lam.

Dohroo, N.P. (2007). Diseases of turmeric. In: Ravindran, P.N., Babu, K.N., and Sivaraman, K. (eds). Turmeric – the genus Curcuma, CRC Press, Boca Raton, USA, pp. 155-168.

Dutta, P., Kakati, N., Das, A., Kaushik, H., Boruah, S., Bhowmick, P., Kaman, P., Puzari, K.C., Bhuyan, R.P. and Hazarika, G. (2017). *Trichoderma pseudokoningii* showed compatibility with certain commonly used inorganic pesticides, fertilizer and sticker cum spreader. *Int. J. Curr. Microbiol. Appl. Sci.*, 6(2): 140-146.

Mahanta, M., Rajesh, T. and Dutta, P. (2021). Isolation and Characterization of the Incitant of Leaf Spot of Turmeric and In-vitro Efficacy of Native Isolate of Endophytic Bacteria. *J. Plant Health Issues*, 2(1):025-029.

Singh, R.K. (2010). Rhizome rot of turmeric. Diseases-Agropedia.

Chapter - 6

Diseases of Coriander (*Coriandrum sativum* L) and their Integrated Management

Apurba Das

Department of Plant Pathology, College of Sericulture,
Assam Agricultural University, Jorhat-785013

Economic Importance

Coriander (*Coriandrum sativum* L.) is a thin stemmed small bushy herb which belongs to Apiaceae family. It is widely used as a seasoning throughout the world. Coriander is commercially cultivated in India, Morocco, Russia, Hungary, Poland, Romania, Bulgaria, Czechoslovakia, Guatemala, Mexico and USA. Approximately 80% of the world's total coriander seed is produced in India. Almost 370 thousand metric tons of coriander have been produced, over an area of 628 thousand hectares in the year 2019-20 (APEDA, 2020). Madhya Pradesh, Rajasthan, Gujarat, Tamil Nadu and Andhra Pradesh are the major coriander growing states of the country. Madhya Pradesh produced the largest volume of coriander. The crop is vulnerable to a number of biotic and abiotic stress which results in a drastic reduction in yield and quality. Coriander is reported to be infected by several plant pathogens at every stage of crop growth. Among the biotic constraints, fungal diseases are the main limiting factor in the production of coriander. In this chapter, we have tried to give a brief account of the major diseases of coriander and its effective management strategies (Champawat and Singh 2008; Das 2014; Tripathi, 2019).

1. Powdery mildew

It is a major fungal disease of coriander which caused 15-40% direct yield loss. Besides this, a considerable volume of coriander seed is additionally lost due to deterioration in quality. It most often starts in mid to late summer. Although the disease usually does not kill the plant outright, however, affects its growth and makes it less productive.

Causal organism

It is a fungal disease caused by *Erysiphe polygoni* D.C. The mycelium of the fungus is well developed with conidia borne in long chains. The foot cells of the conidiophores are cylindrical and straight. The conidia are elliptical to barrel-shaped and measure 25-45x14-26 μm. It produces cleistothecia (measuring 90-135 μm in diameter) with myceloid appendages (Desai, 2002).

Symptoms

Early symptoms of the disease appear as a white powdery mass of fungal spores on the surface of the leaves and tender shoots. In case of severe infection, the entire plant turns white with a powdery coating. Such infected plant parts turn yellow and wither. Early infection prohibits seed formation. Infection in the late stage of the crop results in small and shrivelled seeds, which affect the yield and quality.

Disease cycle

Powdery mildew fungi is an obligate parasite and therefore requires a living host. Year-round availability of host crops or alternate hosts (cucurbits, lettuce, peas, and certain other crops) is important for the survival of the powdery mildew fungi. Resting spores are produced to overwinter in the absence of the main host crop. Powdery mildew spores are spread by wind to new hosts. The spores germinate on the leaf surface and produce mycelium on the undersurface of the leaf. The spores develop a chain of conidia which are attached to fine strands or stalks (conidiophores). The fungal strands become visible as white patches or mildew colonies on the underside of the leaf. Under the favourable condition, the pathogen completes many cycles and lead to severe outbreaks of powdery mildew that economically damage the crop.

Epidemiology

A dry condition is most favourable for the growth and development of the pathogen.

While the spores of powdery mildew fungi are killed and germination is inhibited by the presence of water on plant surfaces for extended periods. Moderate temperatures (16° to 26°C) and shady conditions favour the disease. Spores and fungal growth of the pathogen are sensitive to extreme heat (above 32°C) and direct sunlight.

Management

The disease can be effectively managed by spraying the crop with a 0.2 per cent solution of wet sulphur or Karathane at an interval of 15 days. Dusting the crop with sulphur dust at the rate of 20 to 25 kg per hectare is also effective for checking the disease. Crops grown for seed purposes can be sprayed with two sprays of propiconazole (0.15%) at 15 days intervals (Adiver and Rajanna, 1991, Ali *et al.*, 1999; Sharmila *et al.*, 2004; Vijaya, 2004; Singh, 2006;. Ushamalini and Nakkeeran 2017; Amin *et al.*, 2019).

2. Wilt

The disease attacks the plants at any stage of growth. The severity of the infestation increased with age. As the disease is systemic in nature it is very difficult to control it and results in heavy losses in all coriander growing areas. It causes up to 60 per cent yield loss.

Causal organism

Fusarium oxysporum f.sp. *corianderii*. The fungi produce short monophialides with unicellular, ovoid-elliptical microconidia measuring 3.3-7.6 μm × 1.7-2.8 μm. Macroconidia produced in sporodochia were three-septate and slightly curved. Chlamydospores are terminal and intercalary, which may be single, in pairs, clusters, or in chains measuring 7.9-13.1 μm. (Koike and Gordon, 2005; Gilardi *et al.*, 2019).

Symptoms

Infested plants show drooping of the terminal portion followed by withering and drying of leaves eventually resulting in the death of the plant. Vascular tissue turns brown due to systemic infection of the fungi.

Disease cycle

The pathogen survives as chlamydospores in the soil which serves as the primary source of inoculum. Generally, the disease is a soil born in nature. But in rare cases, the pathogen is found to be seed borne. The fungus can be transported by farm equipment, drainage water, wind, or animals, including humans. Secondary spread occurs through water-dispensable microconidia and macroconidia.

Epidemiology

The pathogen is soilborne and overwinters by producing chlamydospores. High soil moisture and shade favour the high incidence of the disease. A pH range of 5.8 to 6.9 is favourable for maximum incidence. Infection increased with increasing soil moisture (Srivastava 1972).

Management

To prevent the disease use of disease-free seeds is the first and foremost choice. Seed should be obtained from the wilt-free crop/plot of the preceding year. Hot water treatment of the seed lot at 36-40°C for 20 minutes duration is a must before seed sowing. Continuous cultivation of coriander in the same field should be avoided to break the disease cycle. Deep summer ploughing helps in reducing the soil-borne inoculum of the fungi. Soil moisture should be maintained by providing a proper drainage facility. Soil application of 2.5 kg *Trichoderma viride* along with 50 kg FYM is found most effective in suppressing the disease (Manoranjitham *et al.,* 2003; Devappa *et al.,* 2004; Jat *et al.,* 2017)

3. Stem gall disease

It is the most important disease of this crop causing considerable yield loss and deteriorating the quality of seed. Due to the systemic nature of the infection, the disease causes extensive loss (Leharwan and Gupta, 2019).

Causal organism

The fungus responsible for this disease is *Protomyces macrosporus* Unger. The hyphae are intercellular, closely septate and broad, branching is irregular, scattered cells in the hyphae swell, forming ellipsoidal or globose bodies, which later transformed into chlamydospores. As the chlamydospores mature, a thick, hyaline and three-layered wall measuring 50 to 60μm in diameter surrounds them. The mycelium of the fungus is only found in the tumours although the resting spores of the fungus cause systemic infection.

Symptoms

Initial symptoms of the disease can be seen in the form of tumours like swellings on different parts of the plant. Leaf stalks, leaf veins, stems and peduncles are the most vulnerable to the infection. The infected veins show a swollen hanging appearance on the leaves. Initially, the tumours are glossy and rupture later on and become rough. They are about 0.25 inches broad and about 0.5 inches long. In the presence of excessive soil moisture, especially under shaded conditions, when the stem fails to harden and remain succulent the infection becomes most severe. Severely affected plants may be killed.

Disease cycle

The pathogen spreads through soil and seed. It has a complex life cycle including chlamydospores and ascospores. The fungus overwinters in the form of chlamydospores. The spores multiply by budding, as in the case of yeast and spread via air currents. Galls produced after infection serve as the secondary source.

Epidemiology

High soil moisture, high soil pH and soil temperature provide congenial conditions for the rapid growth and development of the pathogen. 22–24°C is the most suitable temperature for chlamydospore germination and disease initiation. Excess soil moisture and low sunshine hours during winter crop enhance the spread of the disease, which may result in total crop failure. Crops sown in mid-October are most vulnerable to the disease.

Management

To minimize the risk of seed-borne inoculum, clean and healthy seeds should be used. Phyto sanitation and crop rotation are mandatory to prevent infection from the soil-borne inoculum. Infected crop residue must be burned after harvest. Seed treatment with suitable fungicide before sowing. Thiram @ 2.5 grams per kg of seed before sowing. Seed treatment and foliar sprays of Hexaconazole (0.2%) after 40, 60 and 75 DAS (0.2%) are the most effective treatment for the management of stem gall (Tripathi, 2005).

4. Damping off or seed rot

Damping off is of serious concern in the young immature stage. Germination to seedling maturity is the susceptible stage for infection. Often more than 50% of plants are killed in the young stage.

Symptoms

Damping off develops in two phases, *viz.*, Pre-emergence damping off and post-emergence damping off. In the first phase, the seedlings failed to emerge from the soil due to infection in the hypocotyls. It results in poor and patchy growth of the seedlings. Post-emergence damping off is characterised by rotting in the collar region. Severely infected seedlings topple down and die.

Causal organism

The disease is incited by a group of fungal pathogens. *Pythium aphanidermatum* and *Rhizoctonia solani* are the most commonly associated pathogen. Both the pathogens perpetuate through soil by producing resting structures viz., oospore and sclerotia. Mycelium of *Pythium* is hyaline, slender, coenocytic and profusely branching. It produces terminal or intercalary sporangia, which are globose to oval. The sporangia germinate to produce the infective zoospores. Towards the end of the growing season, the fungus initiates sexual reproduction. As a result of sexual reproduction, oospores are developed which act as resting spores for the pathogen. The Mycelium of *Rhizoctonia* is septate and multinucleate. The mycelium aggregates to form the sclerotia to survive in the absence of the host.

Disease cycle

Pathogens associated with this disease are seed-borne and soil bore in nature. *P.aphanidermatum* overwinters in the form of mycelium in the plant debris. In certain cases, it also produces oospores in soil which subsequently germinate under favourable weather to produce sporangium to start the primary infection in new crops. *R. solani* mostly survived in soil by producing sclerotia. Hyphae developed from these sclerotia start the infection. The pathogens can survive in the infected field year after year in the form of dormant sclerotia.

Epidemiology

High soil moisture and high relative humidity favour the incidence of damping off. Heavy soil with a pH of more than 6.0 is found to be more favourable for the disease. Pre-emergence damping off is severe at 20-25^{0}C, while a temperature range of 30-35^{0}C favours post-emergence damping off.

Management

The incorporation of sufficient organic amendment in the soil is found effective in reducing the disease incidence. Seed treatment with fungicides like Captaf, captan, and Bavistin at the rate of 0.2% eradicates the seed-borne pathogens. Seed dressing or soil application of *Trichoderma viride* or *Trichoderma harzianum* provide effective control of the disease. Soil drenching with copper oxychloride at the rate of 0.25% protects the crop from soil-borne pathogens.

5. Alternaria leaf blight

Under favourable conditions, the disease is capable of causing the complete crop to lose. Premature loss or collapse of foliage results in significant yield reductions in coriander fields due to the impairment of photosynthesis.

Symptoms

The disease appears as small dark brown to black lesions along the leaf margin of the older lowest leaves. The number and size of the lesions gradually increase, resulting in the browning, shrivelling and scorching of the leaflets. Older leaves are more susceptible than younger ones. In severe cases, the petiole or leaf stems also get infected and develop brown irregular-shaped lesions.

Causal organism

It is generally caused by the fungus *Alternaria poonensis*. But other species of the fungi like *A. alternata* are also found to be associated with the disease in certain instances. The mycelium of both species is septate and profusely branched. The light brown-coloured hyphae become darker when it gets matured. The hyphae emerging through the stromata give rise to the straight or curved conidiophores. Conidia of the fungus are obclavate to pyriform measuring 22.5 - 48.8 µm to 6.55-13.79 µm in size.

Disease cycle

The fungus perpetuates on the seed coat as spores and within the seed as dormant mycelium. Seed borne inoculums are important in the spread of this disease to new areas (Mangwende *et al.*, 2018). The pathogen also survives in the soil both in association with plant residues and as free spores. It can persist in soil for up to 8 years. Leaf blight spores are spread by water, wind and machinery. The spores may come from other diseased fields or from the debris of decomposing leaves.

Epidemiology

It occurs mainly during wet weather in summer, with prolonged heavy dews frequently promoting severe outbreaks in some areas. The disease is favoured by warm and rainy weather and/or overhead irrigation. The optimum temperature for growth and infection is 28°C with some infections occurring at temperatures as low as 14°C and as high as 35°C. Production and transmission of the fungi are increased during moderate to warm temperatures and extended periods of leaf wetness due to rainfall, dew, or sprinkler irrigation.

Management

Seeds for coriander cultivation should be disease free. Seeds should be tested for the presence of pathogens by the floating method. Light seeds should be discarded. Prior to sowing, seeds should be exposed to hot water treatment at 50°C temperature for 20 minutes. Foliar application of Trichoderma-based bio-formulation reduces the disease incidence. Application of PGPR like *Pseudomonas fluorescens* and *Bacillus* spp are found to be effective in suppressing the disease. Seed treatment with captan or Thiram at the rate of 0.2% act as a protectant against the seed-borne fungal spores. Alternating foliar application of Chlorothalonil and mancozeb gives good results for the management of the disease.

6. Stem rot

It is a minor disease of coriander, which is mostly found in cool and humid areas. But under favourable conditions, it may cause considerable yield loss.

Symptoms

The disease is characterised by the presence of cottony, white mycelium that appears on the infected plant surface. Gradually the white fluffy mycelial mat is converted to black-colored sclerotia. The leaves of the infected plants turn yellow and start withering. The severely infected stems get damaged and finally collapsed.

Causal organism

Stem rot is caused by the fungus *Sclerotinia sclerotiorum* (Lib) de Bary. It produces cottony white mycelium over the infected plant parts. Septate, branched and multinucleate hyphae germinate from the dormant sclerotia. Under unfavourable conditions, the mycelium aggregates to form sclerotia measuring 3-4 mm in size.

Carpogenic germination and production of apothecia are the characteristic features of this fungi. On germination of sclerotia, apothecia produce ascospores. The ascospores spread through wind and initiate the primary infection.

Disease cycle

Primary infection is initiated by mycelium arising from myceliogenically germinating sclerotia in soil or by airborne ascospores of *S. sclerotiorum*. Then the fungi produce sclerotia which act as the resting structure during the off-season. It can survive for several years in the soil. It gets germinates under high humidity to form small saucer-shaped structures known as apothecia. White mycelial filaments which grow in the organic matter of soil help in short distance spread of the field within the field. Overhead irrigation may increase disease incidence.

Epidemiology

Sclerotinia is most active when soil temperatures are 12°C to 25°C. Moist soils are necessary for fungal activity. However, once the infection is established, moisture from the plant tissue is sufficient to maintain fungal growth. With high moisture and humidity, the disease spread at an accelerated rate.

Management

Coriander should preferably not be cultivated in fields with a known history of the disease. Deep ploughing of the infected field is necessary to invert the soil to a depth of 250 mm or more to hinder the germination of the sclerotia and hastens their decomposition through antagonistic soil microbes. Severely infected crop fields should be flooded to suppress the pathogen is found effective. The use of plastic mulch in the field gives effective results. Field infection can be successfully controlled by soil application of *Trichoderma*-based bioformulation along with organic amendments like FYM or compost or mustard oil cake. Bio-fumigation with brassica crops suppresses the pathogen inoculum in the field. Spraying or soil drenching with 0.2 % Bavistin can control the disease in the main field.

Conclusion

There is wide acceptability for expanding the area under coriander seed production in India due to its existing productivity and international export potential. Particularly there is an intensive growing demand for organic seed spices with an average annual growth rate of 10 to 20 per cent. Even after the introduction of improved production technologies for coriander, its production is affected by various plant diseases. To resolve the issues more target-oriented research is the need of the hour.

References

Adiver, S. S. and Rajanna, K. M. (1991). Control of powdery mildew of coriander. *Curr. Res.* 20: 59.

Aishwath O.P., Singh H.R., Velmurugan A. and Anwer M.M. (2011). Analysis of soil suitability evaluation for major seed spices in semi-arid regions of Rajasthan using geographic information system, *International Journal of Seed Spices*, 1(1): 29-37.

Ali, S. A., Saraf, R. K. and Pathak, R. K. (1999). Efficacy of fungicides in controlling powdery mildew of coriander (*Coriandrum sativum* L.). *J. Soils Crops* 9: 266–267.

Amin, A.M., Patel, N.R., Patel, J.R. and Amin, A.U.(2019). Management of coriander powdery mildew through new generation molecules. *International J. Seed Spices.* 9(2):96-98.

Champawat RS, Singh V.(2008). Seed spices, disease Management in Arid Land Crops, *Scientific publishers Jodhpur* (India), pp.197-232.

Das, P.C. (2014). Common diseases of spices and condiments: Plant Diseases. Kalyani Publishers, New Delhi.Pp.229-249.

Desai, V.K. (2002). Powdery mildew of coriander caused by *Erysiphe polygoni* D.C. and its management. M.Sc(Agri.) Thesis, Ananda Agricultural University, Gujrat.

Devappa, V., Venkatashalu., Chavan, M.L. and Shantappa, T.(2004). Management of wilt of coriander caused by *Fusarium oxysporum* f. Sp. *Coriandri. The Karnataka J. Hort* .1(1): 61-65.

Gilardi, G., Franco-Ortega, S., Gullino, M. L. and Garibaldi, A.(2019). First Report of Fusarium Wilt of Coriander (*Coriandrum sativum*) Caused by Fusarium oxysporum in Italy. *American Phytopathology.* doi.org/10.1094/PDIS-10-18-1822-PDN.

Jat, M.K., Ahir, R.R., Choudhary, S. and Kakaraliya, G.L. (2017). Management of Coriander Wilt (*Fusarium oxysporium*) through Cultural Practices as Organic Amendments and Date of Sowing. *Int.J.Curr.Microbiol.App.Sci* 6(9): 896-900.

Koike, S.T., and Gordon, T.R. (2005). First report of fusarium wilt of cilantro caused by *Fusarium oxysporum* in California. *Plant Disease.* 89:1130.

Leharwan, M. and Gupta, M.(2019). Stem Gall of Coriander: A Review. *Agricultural Reviews.* 40:121-128.

Mangwende, E., Kritzinger, Q., Truter, M. and Aveling, T. A. S. (2018). *Alternaria alternata*: A new seed-transmitted disease of coriander in South Africa. *European Journal of Plant Pathology.* 152: 409–416.

Manoranjitham, S.K., Rabindram, R. and Doraiswamy, S. (2003). Management of seed borne pathogens and wilt disease of coriander. *Madras Agricultural Journa.l* 90: 4-6.

Sharmila, A. S., Kachpur, M. R. and Patil, M. S. (2004). Field evaluation of fungicides against powdery mildew of chilli (Capsicum annum L.). *J. Mycol. Plant Pathol.* 34: 98.

Singh, A. K. (2006). Evaluation of fungicides for the control of powdery mildew disease in coriander (*Coriandrum sativum* L.). *J. Spices Arom. Crops.* 15: 123–124.

Srivastava, U.S.(1972). Effect of interaction of factors on wilt of coriander caused by *Fusarium oxysporum* f. sp. *Corianderii*, Kulkarni, Nikam & Joshi. *Indian J. Agric. Sci.* 42:618-620.

Tripathi, A.K. (2005). Efficacy of fungicides and plant products against stem gall disease of coriander. *J. Myc. Pl. Path.* 35:388-389.

Tripathi, D.P.(2019). Diseases of spices. In: Crop diseases. Published by Kalyani Publishers, New Delhi. Pp: 246-252.

Ushamalini, C. and Nakkeeran, S.(2017). Studies on management of powdery mildew in coriander using new generation fungicides. *Journal of Spices and Aromatic Crops.* 26 (1) : 59-62.

Vijaya, M. (2004). Chemical control of powdery mildew of okra. *J. Mycol. Plant Pathol.* 34: 604.

Chapter - 7

Diseases of Areca nut (*Areca catechu*) and their Integrated Management

Arti Kumari[1], Ankita Das[2], Madhusmita Mahanta[1]
G. V. Akhila[3] and Pranab Dutta[1]

[1]School of Crop Protection, College of Post Graduate Studies in Agricultural Sciences, Central Agricultural University (Imphal), Umiam, Meghalaya - 793103
[2]Department of Plant Pathology, Assam Agricultural University, Jorhat - 785013
[3]Department of Agriculture Development and Farmer's welfare, Govt. of Kerela

Arecanut (*Areca catechu*) belonging to the family *Arecaceae* is an important tropical cash crop of India, popularly known as betel nut or supari. It is commonly used for mastication along with betel leaves as a mouth freshener. Arecanut is believed to be originated in Malaysia or the Philippines. The crop is grown in tropical countries throughout the globe such as West Indies, Africa, China, India, Bangladesh, Srilanka, Malaysia, Vietnam and other Southeast Asian countries (Anonymous 2014). India is the largest producer of betel vine with total production amounting to 901 thousand metric tons in the year 2019. In India, as of 2019-20, Karnataka stands first in terms of area (2.18 ha) and production (4.57 lt) followed by Kerela, Assam, Meghalaya, West Bengal, Mizoram, Maharashtra and Andaman and Nicobar islands. Thus, it is an important cash crop in the Eastern Ghats, Western Ghats and North-eastern regions of India. Arecanut crop is an integral component of the social, religious cultural and economic life of Indian people. The crop is primarily cultivated for the kernels obtained from fruits which are masticated in tender, ripe or processed form. It is also used in various ayurvedic and veterinary medicines. The use of betel vine dates back to the pre-Vedic period described as taamboola in ancient Indian literature. It is used in various religious ceremonies such as marriage, birth, and puja and is offered to guests as a mark of respect and hospitability.

The areca nut is found to contain various bioactive molecules possessing antioxidant, anti-inflammatory, antimicrobial, antidepressant, anti-hypertension, antiallergic and antidiabetic properties as well as known to protect our body against cardiovascular diseases. The prime constituents include polyphenols, polysaccharides, fats, protein and fibre. Epicatechin, catechin and leucocyanidin are the major polyphenols present in nuts of the betel vine. Although alkaloid content is relatively less but is a significant constituent. The most essential alkaloid responsible for its effect on the central nervous system is arecoline.

Arecanut is a monoecious palm and is capable of growing in a wide range of soil and climatic conditions. It can grow from sea level up to an altitude of 1000 meters and grows well in areas having abundant rainfall or under irrigated conditions. However, the major hindrance in achieving the potential yield of areca nut is the occurrence of a number of pests and diseases. The major diseases of betel nut include Kolerago, bud rot, button shedding, foot rot, band disease, nut splitting and yellow leaf diseases (Chowdappa *et al.*, 2016).

Diseases of Areca nut

1. Koleroga/ Mahali /fruit rot or bud rot

Symptoms

The disease is characterized by extensive shedding and rotting of immature nuts which can be seen lying around the base of the tree. The disease is initially manifested as water-soaked, dark green or yellowish lesions on the surface of fruit near the perianth region. The lesions gradually spread over the fruit surface which leads to shedding and rotting. The infected fruit loses its lustre and quality. The fallen nuts get completely covered with white mycelial growth. With the advancement of the disease, the axis of inflorescence and fruit stalks rot and dry. The infected nuts possess numerous vacuoles and are lighter in weight. The late infection leads to the drying and rotting of nuts without shedding which is known as 'Dry Mahali'.

Causal organism: *Phytophthora meadii*

Disease cycle

The infected bunches of fruit produced at the end of the rainy season may get mummified on the palm and serves as the primary source of inoculum for crown rot, bud rot or fruit rot. The secondary spread of the disease occurs through rain splashes and wind.

Epidemiology

The occurrence and severity of the disease are related to the rainfall pattern of the area. The disease initiates 15-20 post onset of monsoon rains and continues throughout the rainy season. The outbreak of the disease is favoured by low temperature (20 to 23⁰C), and high relative humidity (90%) coupled with intermittent rain and sunshine hours.

Integrated management

Fruit rot can be effectively managed by following an integrated management approach. Field sanitation practices such as the removal and destruction of infected fruit nuts should be followed. Before the onset of the monsoon, a prophylactic spray of Bordeaux mixture @ 1% should be done thrice at 45 days interval. The number of sprays must be increased during prolonged monsoon. Adhesive stickers must be used to ensure the tenacity of the sprayed material on the substrate. Covering the fruit bunch using polythene cover before the onset of monsoon provides complete control.

2. Inflorescence dieback

Symptoms

Disease symptoms initially develop on the rachillae of male flowers, then infects the main rachis with brown-coloured patches which gradually spreads from the tip downwards and later cover the entire rachis leading to the wilting symptom. The female flowers shed from the infected rachis leading to dieback of the entire inflorescence. The discoloured patches harbour fruiting bodies of the fungal pathogen which appear as concentric rings of light pink coloured. Button shedding and inflorescence die-back is a severe problems during monsoon season.

Causal organism: *Colletotrichum gloeosporioides*

Disease cycle

The pathogen survives on infected plant parts and debris as dormant mycelium. The secondary spread of the disease occurs through wind-borne conidia.

Epidemiology

The disease occurs throughout the year but occurs in a severe form during dry conditions i.e. during the month of February-may. Optimum temperature (28-32⁰C) and relative humidity (90-92%) favours disease development.

Integrated management

The disease can be effectively managed by the collection and destruction of infected plant parts which helps in the reduction of the inoculum load of the pathogen. Summer irrigation and optimum utilization of NPK fertilizers help in the reduction of disease severity. Spraying with 0.25% Copper-oxychloride results in the effective management of the disease.

3. Bud rot and crown rot

Symptoms

Bud Rot

The characteristic symptom of bud rot includes chlorosis of the spindle leaf and rotting of the developing bud which can be pulled off easily. The disease progressed to the adjacent leaves showing yellowing and drooping symptoms which shed off completely leaving the bare stem. The infected portion gets attacked by secondary pathogens which transform it into a slimy mass emitting an offensive odour.

Crown rot

Crown rot is characterized by green drooping leaves succeeded by chlorosis of leaf sheath and leaves of outermost whorl thereby leaving spindle leaf healthy, unlike bud rot. Water-soaked lesions appear on the inner portion of the leaf sheath affected by the pathogen. The spear leaf sustains green colour until the bud portion is completely damaged. Later as the disease progress, leaves develop yellow colour, wither and dry up but remain adhered to the stem. Discolouration occurs in the inner tissues of the stem and develops rotting symptoms. At the site of infection, the top portion of the stem gets killed completely.

Causal organism: *Phytophthora meadii*

Disease cycle

The fungal pathogen survives in soil and soil act as the primary source of inoculum. Secondary spread of the disease occurs by dispersal of conidia through rain splash or wind.

Epidemiology

The disease appears with the onset of the southwest monsoon but continues even after the monsoon ceases. During the winter season, occasional rainfall with cooler nights and dew fall favours the growth of the pathogen.

Integrated management

Field sanitation practices such as removal followed by the destruction of infected crowns, fruit bunches and shed nuts reduce the inoculum load of the pathogen. The crown should be drenched with 1% Bordeaux mixture and smearing of 10% Bordeaux paste after removal of infected plant parts are effective in rescuing bud rot infected areca nut palms. To reduce the crown rot incidence, the rhizospheric region should be drenched with the salt of phosphorous acid or potassium phosphonate. Regular monitoring plays a key role in checking the spread of the disease.

4. Anaberoga/ Foot rot disease

This disease causes a loss of about 4.71 % annually.

Symptoms

At the onset of the disease, the leaves of the outer whorl start yellowing and eventually this extends to the inner whorl of leaves. Ultimately, the entire leaves wilt, dry up and fall off leaving the base stem all alone. The stem becomes frail and easily shattered by heavy wind. At the base of the stem, brown spots appear that merge to form bigger lesions and dark fluid also oozes out of the stem. Bracket-shaped fruiting bodies called Anabe are formed at the base of the stem. The roots become rotten and discoloured. The infected trunk when cut open shows brown discolouration up to 1 meter from the ground (Kumar *et al.*, 1990).

Casual organism: *Ganoderma lucidum.*

Disease cycle

The pathogen overwinters in soil and plant debris. The pathogen spreads through wind and splashing water to the healthy plants where they get entry inside the plant through natural openings or wounds and cause infection. The secondary spread of the disease is through airborne spores and infected plant tissues.

Epidemiology

The disease is favoured in neglected plantations with poor drainage facilities and high population density along with acidic hard loamy soil rich in iron and calcium content.

Management approaches

The best management practices for controlling the disease are as follows:

» Removal and destruction of diseased palms.

» Following clean management practices.

» Overcrowding of plants should be avoided.

» Avoid excessive irrigation to reduce disease spread.

» Application of Neem cake @ 2kg/palm/year.

» Making trenches of 30 cm in width and 60 cm in depth around the diseased plant to avoid root contact with a healthy plant.

» Applying Bordeaux mixture @ 1% in the soil before planting.

» Biocontrol agents like *Trichoderma koningi, T. viride, Bacillus amyloliquefaciens, B. subtilis* also help to reduce the disease (Palanna *et al.*, 2017).

5. Yellow leaf disease

An important disease of areca nut causing a yield loss of about 50% over a period of three years after its inception/occurrence.

Symptoms

As the name suggests, the most characteristic symptom of this disease is the yellowing of the leaves starting from the tips of the outer leaves and extending to the middle of the laminae. Brown necrotic lesions also appear parallel to the veins in affected leaves. As the disease progresses, entire leaves of the affected plants turn yellow, dry and wither off leaving a bare trunk. The roots of the affected plants turn black in colour and ultimately rot. The nut quality deteriorates as it becomes soft, and spongy in texture and develops blackish discolouration.

Casual organism

This disease is caused by Phytoplasma

Disease cycle

This disease is transmitted by Plant hopper (*Proutista moesta*) from diseased plants to healthy plants.

Management approaches

The effective way to combat this disease are:

» Use clean, virus-free seeds at the time of planting.

» Rouging off all the diseased plants and debris.

» Planting of resistant varieties like True Mangala and South Kanara.

» Appropriate drainage facilities should be provided to evade water stagnation.

» Potassium and magnesium should be applied in more amounts than the recommended level.

6. Bacterial leaf stripe

This menace of areca nut palm causes a loss of about 60% in seedlings.

Symptoms

The disease initiates as water-soaked, dark green, translucent, linear stripes (1-4 mm wide) running parallel and alongside the midribs of the leaflet. These lesions are usually straight but at times become wavy. These lesions generally emerge from the base or tip of the leaflet. These lesions are covered with slimy, creamy white bacterial exudates on the underside of the leaves and on drying these exudates develop into waxy film or yellowish flakes or fine granules or irregular yellowish masses. As the disease progresses, there may be complete or partial blighting of the leaves thereby killing the entire crown (Naik *et al.*, 2018).

Casual organism: *Xanthomonas campestris* pv. *arecae.*

Disease cycle

The pathogen overwinters in the infected crop debris which serves as the primary inoculum of this disease. When favourable conditions coincide with the pathogen i.e. during the rainy seasons, the bacteria gain entry inside the host through the natural openings on the leaves where they germinate and cause infection. Later on, the pathogen spreads from diseased plants to healthy plants via splashing water, improper intercultural operations and contaminated tools.

Epidemiology

The pathogen requires a temperature of 26 to 28°C, relative humidity of 85-90%, intermittent rainfall, and a susceptible host for causing the disease.

Management approaches

Disease severity can be reduced by the following methods:

> » Removal of diseased plants and its residues from the growing area.

> » Use of healthy planting materials.

> » Closer planting should be avoided along with frequent irrigation.

> » Streptocycline @ 0.05% or copper oxychloride @ 0.3% should be sprayed to control the disease.

> » Biocontrolol agents such as *Trichoderma harzianum*, *Pseudomonas flourescens* can also be used to combat this disease.

7. Leaf blight

In 1964, leaf blight disease in India was first reported by Rao. Earlier, the disease was considered of minor importance but now it has become a serious problem in areca nut growing areas, especially during the rainy season (Nampoothiri, 2000).

Symptoms

The characteristic symptom of the disease includes the appearance of small, round, brown to black coloured spots surrounded by a yellow halo. With the advancement of the disease, several spots coalesce to develop blight. Drying, withering and shredding of leaves may occur in case of serious infection.

Causal organism: *Colletotrichum gloeosporioides* and *Phyllosticta* spp.

Disease cycle

The pathogen survives on infected plant debris for up to 8 months. It can also infect other host plants found in the vicinity of the areca nut plantation. The secondary spread of the pathogen occurs through wind-driven conidia.

Epidemiology

The disease occurs mainly during the rainy season. High rainfall with very high relative humidity favours the outbreak of the disease.

Integrated management

The disease can be effectively managed using an integrated management approach.

Field sanitation must be assured by the removal and destruction of infected plant parts. Plant debris of previous crops should be removed from the plantation field. Chemical control measures include spraying with 0.2% Foltaf for effective management of the disease (Hedge, 2018).

8. Collar rot

Symptoms

Collar rot disease appears in the seedling stage of the crop and mainly occurs in field-planted seedlings or secondary nurseries. Infection occurs at the collar region near soil or root leading to rotting of the developing bud, while infection in the roots causes seedling wilt and death.

Causal organism: *Fusarium* spp. and *Rhizoctonia* spp

Disease cycle

The pathogen survives in soil and infected plant debris which serves as a source of primary inoculums. The secondary spread of the disease occurs through irrigation water and rain splash.

Epidemiology

The disease develops mainly in nurseries. High temperature and relative humidity favour the survival of the pathogen.

Integrated management

The soil-borne pathogens can be managed by maintaining good drainage facilities. Soil drenching with a 1% Bordeaux mixture is also effective in reducing the disease incidence.

Disorders

1. Nut splitting

Nut splitting is a physiological disorder of universal occurrence in almost all orchards. The splitting occurs as the growth of the pericarp fails to keep pace with the developing kernel, thus causing the splitting of the pericarp and distal end. The split nuts drop off prematurely. The split nuts become useless and get secondarily infected by bacteria and fungi.

Symptoms

The symptom initially appears as premature yellowing of the fruit nuts when they are half to three-fourths mature. Yellowing of nuts is followed by splitting into either sides or tips which may extend longitudinally towards the calyx thereby exposing the kernel. It occurs in patches in the orchard and is most common on young palms.

Cause

The splitting of nuts is mainly caused due to excessive flow of cell sap in the inflorescence or excessive supply of nutrients within the inflorescence. Improper drainage and prolonged drought followed by heavy irrigation are the main reasons for nut splitting. The palms of 10-25 years old are most susceptible to this disorder.

Management

The disorder can be managed by the application of Borax @ 0.20% on fruit bunches during the onset of the disease and the application of K_2O fertilizers at the base checks nut-splitting to a great extent. Improved drainage practices and regular irrigation during drought condition helps in the mitigation of the disorder.

2. Hidimundige or Band disorder

Symptoms

The symptoms include the development of crinkled, small and dark green leaves with reduced internodal length and tapering stem. Roots are also poorly developed and appear short, malformed and brittle. Bending of crowns, choking and appearance of oblique rings also become evident on affected palms.

Cause

This physiological disorder is caused due to improper drainage, application of imbalanced nutrients and poor management practices.

Management

The disorder can be effectively managed by improving drainage and adopting proper soil management practices. The hard pan formed in sub-soil should be removed and micronutrients should be applied in appropriate dosages. Organic and inorganic fertilizers should be applied in recommended amount.

3. Sun Scorching

It is a physiological disorder of areca nut caused due to adverse effects of solar radiation.

Symptoms

In the beginning, golden yellow spots appear on the exposed portion of the trunk and soon after these spots turn brown in colour which later on develops into fissures. Furthermore, saprophytes and some insects colonize these infected portions causing more damage to the plants and such palms break away during heavy wind.

Cause

The palms facing the southwestern solar radiation develop a scorching effect.

Management approaches

The best management practices for controlli ng the disease are as follows:

» Planting the seedlings in the north-south direction to avoid direct exposure to solar radiation.

» Growing shade trees like Kokum (*Garcinia indica*), and Banana (*Musa* spp.) on the southwestern side of the garden helps to reduce the scorching effect.

» Wrapping the exposed portion of the trunk with sheaths of areca nut to minimize sun exposure.

» The exposed portion of the trunk is whitewashed/ painted with lime to reduce the disease severity.

Conclusion

Areca nut palm production is affected by a number of diseases and physiological disorders which incur huge economic losses to the crop. Some of the economically important diseases of areca nut include Kolerago, bud rot, button shedding, foot rot, band disease, and yellow leaf diseases. The important physiological disorders affecting areca nut include nut splitting, sun-scorching and band disorder. Most of the diseases occur in a severe form during the monsoon period. Diseases reduce nut yield and quality. The minor diseases include bacterial leaf stripe and leaf blight. The menace caused by the biotic and abiotic malfunctions can be controlled by adopting a blend of management practices like field sanitation, cultural practices, biological control agents and the application of safer chemicals and botanicals. However, extensive research is still needed to develop novel, safe and effective alternatives that will curb the disease and trigger the host defence mechanisms.

References

Anonymous. (2014). AESA-based IPM package, Department of Agriculture and Cooperation, Ministry of Agriculture, Government of India.

Chowdappa, P., Hegde, V., Pandian, R. T. P. and Chaithra, M. (2016). Arecanut diseases and their management.*Indian J. Arecanut, Spices, Med. Plant.*, 18 (4): 46-51.

Hedge, G. M. (2018). Field efficacy of fungicides to manage leaf spot of areca nut. *Adv. Plants Agric. Res.*, 8(6): 496-498.

Kumar, S. N. S. and Nambiar, K. K. N. (1990). Ganoderma disease of areca nut palm: isolation, pathogenicity and control.*J. Plant Crops.*, 18(1): 14-18.

Nampoothiri, K. U. K (2000). Areca nut yellow leaf disease. Kasargod (M) India CPCRI. pp. 15-19.

Naik, S., Gangadhara, N. B., Patil, B. And Revathi, R. (2018). Biochemical characterization of *Xanthomonas campestris* pv. *arecae* causing leaf stripe in arecanut. *J. Pharmacogn. Phytochem.*, 7(1): 1168-1170.

Naik, S., Gangadhara, N. B., Patil, B. and Revathi, R. M. (2018). *In-vitro* evaluation of different chemicals, botanicals and bio agents against *Xanthomonas campestris* pv. *arecae*. *Int. J. Chem. Stud.*, 6(1): 1303-1306.

Palanna, K. B., Narendrappa, T., Basavaraj, S. and Shreenivasa, K. R. (2017). Efficacy of fungal and bacterial bio-control agents on *Ganoderma* spp. causing foot rot of arecanut. *Int. J. Agric. Innov. Res.*, 6(2): 299-304.

Chapter - 8

Diseases of Betel vine (*Piper betle* L.) and their Integrated Management

Pranab Dutta, Anwesha Sharma and Madhusmita Mahanta

Department of Plant Pathology,
College of Post Graduate Studies in Agricultural Sciences,
CAU(I), Umiam, Meghalaya - 793104

Betel vine (*Piper betle* L., Family: Piperaceae), a perennial climber cultivated for its leaf is grown mostly as a cash crop in West Bengal and southern parts of the country, Andhra Pradesh, Telangana, Karnataka, Kerala, Tamil Nadu and to some extent in Bihar, Assam, Orissa, Uttar Pradesh and Tripura. West Bengal contributes about 66% of the total production of Rs. 9000 million.

Several diseases and pests pose threat to the cultivation of this crop. Fungal pathogens like *Curvularia, Phytophthora, Colletotrichum, Fusarium, Sclerotium* etc. contribute the most (Satyagopal *et al.*, 2015; TNAU, 2016; Garain *et al.*, 2020). It is reported that 5-90% of crop loss is due to *Phytophthora* (Dasgupta *et. al.*, 2008), 30-100% loss due to *Colletotrichum* and *Xanthomonas* (Maity and Sen, 1979) and 25-90% loss due to *Sclerotium* diseases (Maity and Sen, 1982).

1. Foot rot or Leaf rot or Wilt

Symptoms

The vines are attacked at all stages of crop growth. Initial symptoms are sudden wilting of vines which show certain yellowing and drooping of leaves from tip to downwards. Leaves turn dull due to loss of lustre. The diseased plant dries up completely within 2 to 3 days. The succulent stem turns brown, brittle and dry. The lower portion of the stem shows irregular black lesions up to the second or third internode. Diseased internodes undergo 'wet rot' and the tissue becomes soft, and slimy releasing a fishy odour. Extensive discolouration and rotting can be seen in the roots of the affected plants.

In young crops, 'leaf rot' symptoms are prominent and the leaves near the soil region show circular to irregular water-soaked spots. The spots enlarge in size and cover a part or whole of the leaf blade, which shows rotting. The leaves turn brown, dark brown or dirty black and defoliation takes place. Leaves within 2-3 feet height of the vine show these symptoms.

Causal organism: *Phytophthora parasitica var. piperina*

Disease cycle

The fungus produces mycelium which is hyaline, non-septate. Sporangia are thin-walled, hyaline, ovate or learn-shaped with papillae, measuring 30-40 X 15-20um. Zoospores, which are liberated from the sporangia, are kidney-shaped and biflagellate whereas the oospores are dark brown, globose and thick-walled.

The fungus survives in infected debris as well as soil. These vines may recover after the rains and survive for more than two seasons till the root infection takes place in collar rot and the death of the vine.

Epidemiology

» Rains from July onwards favour the development of disease.

» September to February months with high atmospheric humidity and low night temperature (below 23°C) is highly favourable.

Integrated management

Cultural

» Removal and destruction of dead vines.

» Planting material should be collected from disease-free gardens and the nursery should be raised and fumigated or solarized.

» Adequate drainage should be provided to stop water stagnation.

»   Injury to the root system during practices like digging should be avoided.

»   The branches of support trees must be pruned at the onset of monsoon to avoid high humidity build-up and for penetration of sunlight.

»   Reduced humidity and the presence of sunlight reduces leaf infection.

»   Irrigation during the cold weather period should be maintained.

Biological

»   Apply *Trichoderma viride, Trichoderma harzianum* and *Pseudomonas fluorescens* as seed and soil application @5g/vine.

»   Apply shade-dried Neem leaves or Calotrophis leaves at 2t/ha and cover this with mud.

Chemical

»   Select matured (more than one-year-old) seed vines from fields and soak them in Streptocycline 500 ppm + Bordeaux mixture 0.05 per cent solution for 30 minutes.

»   Apply 150 kg N/ha/year through Neem cake (75 kg N) and 100 kg P_2O_5 through Super phosphate and 50 kg Muriate of potash in 3 split doses, first dose at 15 days after lifting the vines and second dose and third dose should be given at 40-45 days interval.

»   Copper oxychloride 50% WP @ 1 Kg in 300-400 l of water/acre can also be applied.

2. Sclerotium foot rot and wilt or Collar rot

Symptoms

The vines of all stages are susceptible to this pathogen. Infection usually starts from the collar region. Whitish cottony mycelium can be seen on the stem and roots. Rotting of tissues at the point of attack can be seen at the stem portion and the plants show dropping of leaves and withering finally drying up.

Causal organism: *Sclerotium rolfsii*

Disease cycle

The fungus produces white to grey mycelium with profuse branching. Sclerotia are spherical, smooth and shiny. Brown-coloured mustard-like sclerotia are seen on the infected stem and soil near the vines. The fungus is soil-borne and grows in the dead plant tissue in the soil. The fungus also survives as sclerotia in the infected plant debris in the soil for more than one year. The sclerotia spread through irrigation water. The pathogen can survive on other hosts like chill, groundnut and brinjal.

Epidemiology

May-July months with a high-temperature range of 28-30°C.

Integrated management

Cultural

» Remove the affected vines along with the roots and burn them.

» Application of more of soil amendments like neem cake, mustard cake or farmyard manure is preferable.

» Adequate drainage should be provided to stop water stagnation.

» Injury to the root system during practices like digging should be avoided.

» The branches of support trees must be pruned at the onset of monsoon to avoid high humidity build-up and for penetration of sunlight.

» Reduced humidity and the presence of sunlight reduces leaf infection.

» Irrigation during the cold weather period should be maintained.

Biological

» Apply *Trichoderma viride*, *Trichoderma harzianum* and *Pseudomonas fluorescens* as seed and soil application @2kg/50kg FYM.

» Apply shade-dried Neem leaves or Calotrophis leaves at 2t/ha and cover them with mud.

Chemical

» Drench the soil with 0.1 % Carbendazim.

» Copper oxychloride 50% WP @ 1 Kg in 300-400 l of water/acre is effective control measure.

3. Anthracnose

Symptoms: The leaves show small, circular spots initially, which gradually enlarge and develop to a size of 2 cm in size. The spots are irregular in shape and size, light to dark brown surrounded by a diffuse chlorotic yellow halo. Marginal leaf tissue becomes black, necrotic and spreads towards the centre of the leaf. Occasionally diffused yellow halo also develops. In the later stage circular, black lesions that occur rapidly, increase in size and girdle the stem, resulting in withering and drying and finally death of the vine.

Causal organism: *Colletotrichum piperis*

Disease cycle

The fungus produces a large number of acervuli containing short, hyal conidiophores block-red setae. The conidia are single-celled, hyaline and falcate. The fungus remains in the infected plant debris in the field. Soil-borne conidia serve as primary inoculum, spread by rainwater splash or splash irrigation. The secondary spread in the field is caused by air-borne conidia.

Epidemiology

» High rainfall favours the growth of the pathogen.

» High humidity is responsible for the development of disease.

Integrated management

Cultural

» Select healthy and disease-free planting materials.

» Use resistant and tolerant varieties.

» Infected vines and leaves should be collected and destroyed.

» Eradication of infected vines from the vineyard.

» Apply phytosanitation process and irrigation with rose can.

Biological

» Apply *Trichoderma viride, Trichoderma harzianum* and *Pseudomonas fluorescens* as seed and soil application.

» To boost the crop growth use Neem cake @40 kg/acre under assured moisture conditions.

Chemical

» Spray 0.2 % Ziram or 0.5 % Bordeaux mixture after plucking of the leaves.

» Foliar spray of 0.05% bitertanol, followed by 0.2% Mancozeb, 0.1% Ziram and 0.5% Bordeaux mixture at 20 days interval is highly effective.

» In severe conditions, drench soil with Metalaxyl 18% WP + Mancozeb 64% @2g/litre of water.

4. Fusarium wilt

Symptoms

Plants show the yellowing of leaves and wilting starts gradually. Often sudden drying up takes place in the entire plant. Vascular discolouration of plants can also be seen.

Causal organism: *Fusarium solani*

Disease cycle

Chlamydospores survive in soil. Conidia spread through irrigation water. Secondary infection takes place by conidia through rain or wind.

Epidemiology

High temperature and high relative humidity favour the development of disease in the vines.

Integrated management

Cultural

» Remove the affected vines along with the roots and burn them.

» Application of more of soil amendments like neem cake, mustard cake or farmyard manure is preferable.

» Adequate drainage should be provided to stop water stagnation.

» Injury to the root system during practices like digging should be avoided.

» The branches of support trees must be pruned at the onset of monsoon to avoid high humidity build-up and for penetration of sunlight.

» Reduced humidity and the presence of sunlight reduces leaf infection.

» Irrigation during the cold weather period should be maintained.

Biological

» Soil treatment with *Trichoderma viride* or *T. harzianum* @ 2.5 kg/ha

» Apply shade-dried Neem leaf or Calotrophis leaves at 2t/ha and cover it with mud.

Chemical

» Drench the soil with 0.1% Carbendazim.

» Copper oxychloride 50% WP @ 1 kg/300-400 l of water/acre is preferable.

5. Powdery Mildew

Symptoms

White to light brown powdery patches appear on lower surface of the leaves, which gradually increase in size. Early leaf infection appears as light grey spots which enlarge and turn into a powdery mass of fungal growth covering the lower surface of the leaf. Under ideal conditions, both the upper and lower leaf surface gets covered by the white floury mass of fungal growth resulting in early leaf fall.

Causal organism: *Oidium piperis*

Disease cycle

The fungus is ectophytic and produces profusely branched, hyaline and septate hyphae on the surface of the leaves. Conidiophores are short, club-shaped, non-septate and hyaline and produce conidia in chains. Conidia are single-celled, elliptical, and borne over short conidiophores. The fungus survives in the crop debris. Secondary spread occurs through wind-borne conidia.

Epidemiology

» Cool weather with mild temperature favours the development of disease.

» Dry humid weather during the months of May-July also causes pathogen growth in some conditions.

Integrated management

Cultural

» Remove the affected vines along with the roots and burn them.

» Reduced humidity within the vineyard, good air circulation through the canopy, and good light exposure to all leaves and clusters, aid in managing powdery mildew.

» Use an under-vine irrigation system and manage it carefully without excess to prevent the disease.

Biological

» Apply *Trichoderma hamatum*, *Trichoderma harzianum* and *Trichoderma viride* are the most promising methods.

» *Pseudomonas fluorescens* as seed and soil application.

» Spraying neem oil, and canola oil, reduces powdery mildew severity in the plants.

Chemical

» Spray 0.2 % wettable Sulphur or dust Sulphur @ 25 kg/ha after plucking of the leaves.

» Spray Bavistin @1g/litre of water.

6. Bacterial leaf spot or stem rot

Symptoms

Bacteria produce minute water-soaked lesions that appear all over the leaf blade which is delimited by veins. These coalesce to form large irregular brown spots. Premature defoliation takes place among the infected leaves.

Causal organism: *Xanthomonas campestris* pv. *betlicola*

Disease cycle

The pathogen survives in soil, and bacteria spread through irrigation water. Viable bacteria in the infected vines and leaves serve as a primary source of inoculum. Rain splashes and splash irrigation water help in the secondary spread of the pathogen.

Epidemiology

» High temperature favours the development of disease.

» Cloudy weather with intermittent rains and high humidity favours pathogen growth.

» 2 to 3 years old vines are highly susceptible to attack.

Integrated management

Cultural

» Planting material should be collected from disease-free gardens and the nursery should be raised and fumigated or solarized.

» Collect and burn the infected plant parts to minimise the spread of the disease.

» Increase air circulation in the vineyard.

» Remove disease cane during normal pruning operations in the dormant season.

» Follow up with hand pruning.

Biological

» *Bacillus subtilis* is a successful biocontrol agent of the pathogen.

» *Bacillus cereus,* and *Bacillus megaterium* also effectively control the pathogen attack.

Chemical

» Spray Streptocycline 400 ppm+ Bordeaux mixture 0.25 per cent at 20 days intervals, after plucking the leaves.

» Spray Plantomycin 500 ppm for disease control.

Future Prospects

Good progress has been recorded regarding the disease management for betel vine. Identification of constituents in different landraces using modern scientific techniques could be useful for future elite land races selection and their improvement programmes. Also, efforts should be made on the characterization of the available landraces which could be useful for resolving similar problems and their proper authentication. Proper characterization could be useful for long-term research for drug development due to betel vine's therapeutic value. It is also imperative to study the effect of abiotic factors on the production and quality of betel vine and its relation to disease development. Recent biotechnological tools like chromatography, NMR other functional genomics techniques could be explored to find out new compounds with active potential from this unexplored plant species, develop disease-resistant high yielding varieties, etc. (Das *et al.*, 2016). Special attention is needed in pest and disease management in betel vine for the development of new and improved varieties (Jana, 2017). Further study is needed in chlorophyllase activity for long-term storage and improvement of export potential of a betel vine leaf. Genetic diversity assessment using molecular markers should be intensified taking a maximum number of landraces.

Conclusion

The common serious problem in the cultivation of betel vine is its diseases like powdery mildew, foot rot, and leaf rot caused by various pathogens, causing great loss at both national and international economic levels. The factors influencing the growth of betel vine inside the 'baroj' are low temperature, high humidity and diffused light. The interrelationship between the climatic factors and the frequency of diseases should be studied before giving a final recommendation and suggestion to the farmers. In essence, the biotechnological intervention has opened up a new horizon for genetic improvement including the development of major disease-resistant varieties, proper authentication and identification of elite chemotypes of this medicinally and economically important cash crop.

References

Das, S., Parida, R., Sandeep, I. S., Nayak, S. and Mohanty, S. (2016) Biotechnological intervention in betelvine (Piper betle L.): A review on recent advances and future prospects. *Asian Pac. J. Trop. Med.*, 9(10): 938-946.

Dasgupta, B., Mohanty, B., Dutta, P. K., and Maiti, S. (2008). Phytophthora Diseases of Betelvine (*Piper betle* L.): A Menance to Betelvine crop. *SMRC J. Agric.*, 6: 71-89.

Garain, P. K., Mondal, B., Maji, A. and Dutta, S. (2020) Survey of Major Diseases in Mitha Pata variety of Betelvine (Piper betle L.) under Coastal Saline Zone of West Bengal, India. *Int. J. Curr. Microbiol. Appl. Sci.*, 9(3): 2490-2498.

Jana, H. (2017) Plant Protection Measures to Control Insect-Pest and Diseases of Betel vine. *Rashtriya Krishi*, 12(1): 17-20.

Maiti, S. and Sen, C. (1979). Fungal diseases of Betelvine. *PANS*, 25: 150-57.

Maiti, S. and Sen, C. (1982). Incidence of major diseases of betel vine in relation to weather, *Indian Phytopathol.*, 35(1):14-17

Satyagopal, K., Sushil, S.N., Jeyakumar, P., Shankar, G., Sharma, O.P., Sain, S.K., Boina, D.R., Rao, N.S., Sunanda, B.S., Asre, R., Murali, R., Arya, S., Kumar, S., Kalra, V.K., Panda, S.K., Sahu, K.C., Mohapatra, S.N., Ganguli, J., Lakpale, N., Sathyanarayana, N. and Latha, S. (2015) AESA based IPM package for Betelvine, pp. 38.

TNAU (2016) Diseases of Betel vine. Diseases of Horticultural Crops & Their Management, pp. 142-146

Chapter - 9

Diseases of Jute (*Corchorus* spp.) and their Integrated Management

Jyotim Gogoi[1], Amar Bahadur[2] and Pranab Dutta[3]

[1&3] School of Crop Protection, College of Post Graduate Studies in Agricultural Sciences, Central Agricultural University (Imphal), Umiam, Meghalaya
[2] Department of Plant Pathology, College of Agriculture, Lembucherra, Tripura

In India jute is the most important fiber crop after cotton. Raw jute fiber is considered a source of raw material for packaging manufacturing processes. But it uses for applications in textile industries, paper industries, building and automotive industries, uses as soil mulch materials etc. the jute fiber is also used in a more refined way to produce cloth and garments. The primary cultivars of jute belong to two species *Corchorus olitorius* but the fiber is superior as compared to *Corchorus capsularis*. The jute crop is cultivated throughout the country all about 8 lakh hectares and productivity is up to 22.12 q/ha including 10.58 million bales of 180 kg each. The major jute-growing states are North Bengal, Bihar and North-Eastern States including Assam, Meghalaya, Nagaland and Tripura. The major disease of jute is Stem rot caused by *Macrophomina phaseolina*. Other diseases include Anthracnose (*Colletotrichum corchorum* and *C. gloeosporioides*), Black band/Die-back (*Botryodiplodia theobromae*), Soft rot (*Sclerotium rolfsii*), Jute Mosaic (begomovirus) and Hooghly Wilt (*Ralstonia solanacearum*, *Meloidogyne incognita*, *Rhizoctonia bataticola* and *Fusarium* complex) (Sarkar *et al.*, 2016; De, 2019).

Table 1: Diseases of jute

Sl. No.	SN Name of disease	Nature	Causal organism
1	Stem rot	Fungal	*Macrophomina phaseolina* (Tassi) Goid.
2	Hooghly wilt	Bacterial	*Ralstonia* (= *Pseudomonas*) *solanacearum* (Smith) Smith (*Macrophomina phaseolina, Meloidogyne incognita* may facilitate entry of bacteria by making injury).
3	Anthracnose	Fungal	*Colletotrichum corchorum* Ikata & Tanaka; *C. gloeosporioides* (Penz.) Penz. and Sacc.
4	Black band	Fungal	*Botryodiplodia theobromae* (Pat.) Griff. and Maubl.
5	Soft rot	Fungal	*Sclerotium rolfsii* Sacc. (=*Athelia rolfsii*)
6	Tip blight	Fungal	*Curvularia subulata* (Nees ex Fr.) Boedijn
7	Stem gall	Fungal	*Physoderma corchori* Lingapa
8	Powdery mildew	Fungal	*Oidium* sp.
9	Sooty mould of pods	Fungal	*Cercospora corchori* Sawada, *Corynespora cassicola* (Berk. & M. *A. Curtis*), *Alternaria* spp.
10	Die back	Fungal	*Diplodia corchori* Syd. & P. Syd.
11	Root- gall nematode	Nematode	*Meloidogyne incognita* (Kofoid and White) Chitwood, *M. javanica* (Treub) Chitwood
12	Jute mosaic	Viral	Begomovirus under family *Geminiviridae,* vector: *Bemisia tabaci* Genn. (Whitefly).
13	Jute Chlorosis	Viral	A member of Tobravirus genus
14	Yellow vein disease	Viral	A bipartite Begomovirus, vector: whitefly (*Bemisia tabaci* Genn.)

(*Source*: Jute Diseases: Diagnosis and Management)

1. Stem rot of jute

Symptom

The pathogen infecting the crop leads to damping off, seedling blight, leaf blight, collar rot, stem rot and root rot.

Damping off

After germination deep brown spots are noticed on the cotyledonary leaves giving a blighted appearance called seedling blight. Under humid conditions, browning reaches the roots and the seedling dies which is called as damping off.

Seedling blight

In the month of June-July when the plants are about more than seventy days old several small brown spots are found on the leaves that gradually increase in size and coalesce with each other to form a bigger brown-coloured rotted area. The disease generally spreads from the infected leaves in two ways viz. the pathogen may enter the stem through the petioles of the infected leaves or the infected leaves fall and adhere to the stem surface where the infection may occur.

Stem rot

The most characteristic symptom of the disease is the formation of blackish brown lesions or depressions on the stem which increase in size and several such lesions may coalesce and finally girdle the stem. All the leaves fall from the infected plant and the stem looks black or dark brown. In the case of seed crops, the late infection may cause spotting on the capsules and the formation of pycnidia and sclerotia on the capsule and seed.

The disease is severe in Assam, West Bengal, Bihar and Orissa reducing the yield and quality loss of fiber. It reduces the total by 10%, but it can go up to 35-40% loss in severe infection.

Causal organism

Macrophomina phaseolina (Tassi) Goid causes stem rot of jute including 72 families. The pycnidia are about 100-200 μm in diameter, dark to greyish, becoming black with age, globose or flattened globose, membranous to subcarbonaceous with an inconspicuous or definite truncate ostiole.

Epidemiology

Acidic soil condition (pH 5.6-6.5), high level of N_2, high intensity of rainfall and RH favours the infection of *Macrophomina phaseolina*. Low moisture content and high soil temperature lead to infection in older plants. Sclerotia germinate in the soil temperature between 28^0C to 35^0C. Germ tubes penetrate the mechanical pressure and enzymatic digestion or through natural openings. Airborne conidia are responsible for the secondary spread of the diseases under favourable conditions viz. cloudy weather, high rainfall and temperature of about $(30-35)^0C$. Sclerotia survive in the soil in crop debris for up to 3 years.

Management

Cultural practices

Continuous cultivation of jute cause a sharp reduction of calcium and potassium leading to increases in soil pH. Acidic soil favours stem rot disease. Application of lime or dolomite @ 2-4 t/h before crop sowing reduces the stem rot diseases. Sowing of the crop in April month reduces the infection with optimum spacing row to row 25-30 cm and plant to plant 5-6 cm because closure spacing increases the infection of the disease and spread of the diseases. Limited use of nitrogenous fertilizers and application of K_2O @ 50-100 kg/ha limits the severity of the diseases. Proper drainage should be necessary in case of tossa jute, which cannot withstand waterlogged conditions.

Chemical control

The seeds are the major source of pathogen inoculum therefore seed treatment with Carbendazim 50 WP @ 2g /kg or Dithane M 45 (Mancozeb) @ 5g/kg provides maximum protection against the stem rot diseases of jute. When infection increases by 2% scheduled application of the following fungicide should be applied Carbendazim 50 WP @ 2g/l, Dithane M 45 @ 5g/l or Copper oxychloride 50 WP @ 5-7 g/l Tebuconazole 25.9 EC @ 0.1%. In severe infection 3-4 sprays should be done at an interval of 15-20 days.

Biological control

Bioagents such as *Trichoderma viride*, *Aspergillus niger* and fluorescent *Pseudomonas* can reduce the diseases. The soil of *Trichoderma viride* enriched FYM/compost reduces the incidence of the disease. Soil inoculant with PGPR such as *fluorescent Pseudomonas*, *Azotobacter* and *Azospirillum* reduces the diseases and gives better quality fiber and yield.

Resistant variety

No jute variety is resistant to stem rot but tossa jute JRO 524 and JRO 632 and white jute cultivars JRC 212 and JRC 321 showed reduced infection. In Assam, the jute variety OIN 125, OIN 154, OIN 651 and OIN 853 are moderately resistant to stem rot.

2. Anthracnose

The disease entered in India through Assam and caused a severe loss in the capsularis belt at regular intervals *viz.*, Assam, North Bengal, Bihar, and Uttar Pradesh.

Symptom

At seedling stage, the pathogen infects on leaf and the stem leading to a brownish spot and streaks followed by drying up of leaves. In later stages of the plants, initially, light yellowish patches are seen on the stem which turns to brown/black depressed spots. The spots on the stem are irregular in shape and size. The spots may coalesce causing deep necrosis showing crakes on the stem and exposing the fibre tissues. Severely infected plants die and fall apart because of the coalescing spots girdle the stem the plants break at that point. Pods of diseased plants are also exaggerated showing depressed spots and seeds collected from such fruits are also infected. The infected seeds are lighter in colour, shrunken and germination is poor.

Causal organism

Colletotrichum corchorum infecting in white jute and *C. gloeosporioides* infecting in tossa jute.

Epidemiology

The pathogen survived in soil, seeds and crop debris. The mycelium of the pathogen entered the epidermis and attacks the parenchymatous tissues. In humid conditions, the acervuli are seen the infected regions with necked eyes. The diseases progress in hot and humid climatic conditions during the month of July when the crop is about two months old. Continuous rain, high relative humidity and temperature of around 35^0C favour fast infection and spreading of the diseases.

Management

Seed treatment with Carbendazim 50 WP @ 2 g/kg or Captan @ 5 g/kg gives better control of the diseases. Spraying the infected crop with Carbendazim 50 WP @ 2 g/l or Captan @ 5 g/l or Mancozeb @ 5 g/l controls the pathogen. The seeds which are infected with more than 15% should not be used for sowing.

3. Black band/Die-back of jute

Symptom

The infection leads to discolouration of the tips of main shoots that precedes gradual darkening and withering of branches which ultimately resemble blackened stocks. Innumerable, erumpent pycnidia, which extrude masses of spores are produced. The infection gradually occurs in mature plants 2-3 feet high from the soil surface the soil on the stem first appears as brownish colour later it becomes a black band. The disease first appears as a small blackish-brown lesion which gradually enlarges and encircles the stem resulting in the withering of epical and side branches.

Causal organism

Botryodiplodia theobromae

Management

Seed treatment with Carbendazim 50 WP @ 2g/kg. Foliar spraying with Carbendazim 50 WP @ 2g/l water or Cooperoxychloride @ 5-7 g/l water or Mancozeb @ 4-5 g/l water gives maximum protection.

4. Soft rot of jute

Symptoms

Infection is initiated from the base of the stem and near the soil surface. The symptom appears like a soft, brown wet patch at the base of the stem. Cottony mycelial growth seen in the near base later transforms into sclerotia. Infected parts become golden brown when the crop is about 80-90 days old.

Causal organism

Sclerotium rolfsii

Mycelia is golden brown in colour and mycelia profusely branched. The pathogen has a wide range of hosts and infected both *capsularis* & *olitorius* jute crops.

Epidemiology

Hot soil and high humidity favour infection of the pathogen. The sclerotia survive in the soil for several years. The sclerotia propagate in the fallen leaves and spread the disease in the ongoing crop.

Management

Cultural practices

Deep ploughing soil during summer months to exposure of hibernating soil sclerotia to sunlight. Application of optimum quantity of FYM/compost. Clean cultivation removing crop debris of the previous crop. Fallow cropping with non-host crops and cultivation of green manure crops like Sunn hemp or *dhaincha*. Soil application of FYM/compost enriched with *Trichoderma viride*.

Chemical control

Spray the crop with Copper oxychloride @ 4-5 g/L of water at the base regions of the plants. Mancozeb 2000 μg/ml also completely inhibits the growth of the *Sclerotium rolfsii* in laboratory conditions.

5. Tip blight

The disease was reported to infect the jute variety JRO 8432 but was previously considered a minor disease of the jute crop.

Symptom

A newly emerged twig of the infected by the pathogen leads to a blighting appearance. The unfolded tip of the plants turns brown, rotten slowly and falls apart in highly humid conditions. The infection is relatively low in low humid conditions.

Causal organism

Curvularia subulata

6. Stem Gall

Symptom

The plants suffer from waterlogged conditions prone to this gall infection near the water line of the stem. The size of the gall varies from 4-10 cm in diameter. The galls appear as lentil seed-like outgrowth on the stem. The gall later turns rusty and becomes black plants wither and die. The disease is widely known as pox of jute. The pathogen only infects the *Olitorius* jute but not on *capsularis* jute.

Causal organism

Physoderma corchori Lingappa

Management

Spraying the crop with Copper oxychloride 50 WP @ 4.0 g/L of water gives maximum protection from diseases.

7. Powdery Mildew

Symptom

The crop both *olitorius* and *capsularies* are infected by the pathogen in dry areas at later stages. The white powdery growth can be seen on leaves and seed coat giving a chalky appearance.

Causal organism

Oidium sp.

Management

Dusting the crop with lime-sulphur in the morning hours gives a good result with practicing clean cultivation.

8. Hooghly wilt

Symptom

Dropping and weathering of leaves of the plants starts from the basal portion of the leaves. The infected stem becomes soft, slimy fluid comes out on slight pressing. The ooze test confirms the wilt is caused by bacteria. In later stages of the disease progress all leaves dropped down only roots and stem remains. The leafless erect is stem visible from long distances.

Causal organism

A complex combination of pathogens caused the Hooghly wilt *viz.*, *Macrophomina phaseolina*, *Pseudomonas solanacearum* and *Fusarium solani* infecting jute followed by potato crop. Because of the complex nature of the causative agents, the disease termed as Hooghly wilt by T. Ghosh originated in the Hooghly district of West Bengal. Later confirmed that the disease is caused by *Ralstonia solanacearum* (=*Pseudomonas solanacearum*) is the original pathogen, whereas *M. phaseolina* and *Meloidogyne incognita* help the bacteria to penetrate the plant cell wall.

Management

Cultural practices

Do not fallow crops with solanaceaous crops such as potato or tomato. Jute: Paddy: Paddy or Jute: Paddy is the best combination of the crop to reduce the Hooghly wilt. At least for 2 years *rabi* crop rotation with non-solanaceous like wheat. Clean cultivation and destruction of wilted plants give better protection.

Chemical control

Seed treatment with Carbendazim 50 WP @ 1.0 g/kg of seeds. Spray the crop with Carbendazim @ 1 g/liter of water to give better control of stem rot.

9. Jute mosaic

(Synonyms: Jute leaf mosaic, Jute yellow mosaic, Jute golden mosaic):

Symptom

In early stages of the crop small yellow irregular dots (flakes) appears on the leaves. Later stage the spots intermingled with green patches of the leaves and creates a yellow mosaic appearance. The infection leads to a reduction in leaf size, chlorotic tissue become yellow and plants look pale.

Causative agent

Begomovirus under family Geminiviridae

Transmission

The virus is transmitted by whitefly (*Bemisia tabaci* Gen.). The seed transmission also is accounted for. Viruliferous whitefly holds the virus inoculum for 4 hours with 10 days cycle.

Management

Rough out the infected plants. Spray the crop with systemic insecticide Imidacloprid @ 2.0-3.0 ml/10liter, Thiomethoxam @ 2.5-3.0gm/10liter and Acetameprid @ 1 g/5liter of water can reduces the spread of the disease. Use of healthy seeds and spray the crop with systemic insecticide in 30 days old crop give batter result. A booster dose of nitrogen after 45-50 DAS is necessary to increase plant immunity.

10. Root knot nematode

Symptom

The nematode caused the formation of globular swellings in roots termed gall. Infected plants suffer from reduced translocation of food and water due to root infections caused by the nematode. The infected plants become yellowish and stunted. Sometimes the infection leads to Hooghly wilt. Because the nematode penetration gives the entry of the bacterium *R. solanacearum* and fungal pathogen *M. phaseolina*.

Causal organism

Meloidogyne incognita (Kofoid and White) Chitwood, *M. javanica* (Treub) Chitwood

Management

Thiometon, Nematox, Nemagon and granular form Carbofuran, Phorate are effective against RKN. Soil amendments *viz*., cakes of karanj, mahua, groundnut, sawdust, cow dung, castor, and chicken manure give a reduction in nematode infection. Clean cultivation and removal of infected plants with a crop rotation of non-host crops like paddy and wheat for at least two years give better protection. Resistant variety like JRO 524 and JRO 632 shows good result in yield and dry matter after infection of nematode J 2 stage.

11. Little leaf and bunchy top

Newly emerging disease of jute crop. The infected plants showed profuse lateral branching with a bushy appearance. The terminal leaves show a bunchy appearance with reduced height. Reported from CRIJAF, Barrackpore in 2012.

Causal organism

Phytoplasma

IPM module for jute

IPM module consisting of cultural with deep ploughing to expose soil to sun, sowing in line include 5-6 lakh plants/ ha, NPK: 60 (30+15+15):30:30, manual hand weeding once at 21 DAS, variety: JRO 204 (Suren)); chemical (soil application of Ca(OCl)$_2$ @ 30 kg/ha at 7 DBS, seed treatment with (a) Carbendazim 50 WP @ 1 g/kg + (b) Imidacloprid @ 4g/kg, application of pesticides such as Spiromecifen @ 1 ml/litre, Profenophos @ 2 ml/litre); biological control agents (seed treatment with *Trichoderma viride* @ 10g/kg, soil application of *Pseudomonas fluorescens* @ 100g/ sq. m before sowing and spraying of neem oil @ 3-4 ml/litre) components is effective against stem rot, yellow mite, semi-looper, bihar hairy caterpillar, apion and indigo caterpillar infestation of jute at institute farm level.

References

Parthasarathy, S. (). Diseases of Jute. Dept. of Plant Pathology, College of Agricultural Technology, Theni http://cattheni.edu.in/wp-content/uploads/2018/09/Diseases-of-Jute-and-Sunnhemp.pdf

De, R.K. (2019). Jute Diseases: Diagnosis and Management, Tech. Bull. No. 04/2019. ICAR- Central Research Institute for Jute and Allied Fibres.

Sarkar, S.K. and Gawande, S.P. (2016). Diseases of Jute and allied fibre crops and their management. *J. Mycopathol. Res.*, 54(3): 321-337.

https://farmer.gov.in/cropstaticsjute.aspx

https://plantlet.org/different-disease-of-jute/

Chapter - 10

Diseases of Mesta (*Hibiscus* spp.) and their Integrated Management

Gitashree Das[1], Jyoti Kumari[2] and Pranab Dutta[3]

[1]Departmnet of Plant Pathology, Assam Agricultural University, Jorhat, Assam
[2] Department of Plant Pathology, Dr Rajendra prasad Central Agricultural University, Bihar
[3] SCP, College of Post Graduate Studies in Agricultural Sciences, Central Agricultural University(Imphal), Umiam, Meghalaya

Introduction

Mesta, a herbaceous yearly plant (ligno-cellulosic bast fiber crop like jute) accepted to be started from Afro-Asian nations, positions close to jute in significance (sharing 15% of crude jute-cum-mesta fiber creation) and involves two significant particular developed species– *Hibiscus cannabinus* (Kenaf, 2n = 36) and *H. sabdariffa* (Roselle, 2n = 72) having a place with cotton family, Malvaceae, request: Malvales.

Mesta is more versatile than jute under assorted states of environment and soil and it is additionally extremely impervious to dry season. It is harder and more grounded than jute fiber yet is to some degree coarser and less graceful. It, nonetheless, rises in quality to the medium grades of jute (Mahapatra *et al.*, 2009).

Mesta has demonstrated as a significant substitute for jute and is effectively being filled in tropical and subtropical areas of both halves of the globe. The head producing countries are India, China, Thailand, Egypt, Sudan, Brazil and Australia. In India, mesta is grown in 13 States namely Andhra Pradesh, Bihar, Chattisgarh, Jharkhand, Karnataka, Madhya Pradesh, Maharashtra, Meghalaya, Odisha, Tamil

Nadu, Tripura and West Bengal.

Some people used young fruit and leaves of mesta as a vegetable. Mesta is used in many industrial purposes making gunny bags, twine, ropes, papers etc. It gives an average yield of 10-13 quintals/ acre. Due to biotic and abiotic stresses, mesta crops are affected by many diseases and pests. Diseases and pests reduce the production and productivity of the crop. Important major diseases are foot and stem rot, stem rot, mesta leaf blight, and yellow vein mosaic virus (Anonymous 2014-15; Chatterjee and Ghosh , 2008; Islam *et al.*, 2013).

1. Foot and stem rot

Foot and stem decay is predominant in all the mesta developing spaces of India for example, West Bengal, Andhra Pradesh, Odisha, and Bihar. It is the most significant illness of mesta in India causing around 10 – 25% misfortune in fiber yield. In extreme cases, over 40% of crop misfortune in roselle was noticed. The pathogen attacks the plant at the very beginning phase and proceeds up to development influencing the quality and amount of fiber. It is more significant in roselle (*H. sabdariffa*) than kenaf (*H. cannabinus*).

Symptoms

The side effect shows up on the stem commonly a couple of creeps over the ground yet the spots might be seen at a sequential level moreover. The spots are profound brown to blackish in shading with variable size. Bigger spots all the time support the stem and as a result, the plant breaks at the place of contamination. No fiber is acquired from such plants.

Causal organism

Foot and stem rot of mesta is caused by *Phytophthora parasitica* var. *sabdariffae*.

Epidemiology

Phytophthora parasitica var. *sabdariffae* is a soil-borne pathogen supported by high temperature (30° - 35°C) and continuous showering. Water stagnation which is common during the mesta developing season in West Bengal inclines the plants to contamination. Ceaseless drizzling, high precipitation and shady condition from May to September might be answerable for a pandemic (De and Mandal 2007b).

Management

None of the developed varieties is impervious to this disease yet AMV 1, Roselle, Type 1 and AP 481 were seen to be moderately safe. Red shuddered *H. sabdariffa* lines are safer. Seed treatment with Dithane M 45 (Mancozeb) @ 5.0 g kg/l followed by soil dressing (0.2%) and splashing of Copper oxychloride 50 WP @ 5.0–7.0 g/l or Carbendazim 50 WP @ 2.0 g/l of water. Pre-planting seed treatment with copper oxychloride was more successful than Carbendazim, Mancozeb, Metalaxyl, Ediphanphos, Carboxim, Hexaconazole and Thiophanate methyl. Copper oxychloride lessens the disease frequency by 50.3% and 45.5% at 30 days subsequent to planting and at development separately and expanded 20.6% fibre yield.(De and Mandal 2007a).

2. Stem rot (*Sclerotinia sclerotiorum*)

Stem decay of kenaf and roselle was a minor disease however during the most recent two years (2012-14). The incidence of the disease was just about as high as 50 % in seed crops at CRIJAF Farm, Barrackpore. The contaminated plant neither yields any fiber nor seeds.

Symptoms

The sickness shows up as water-doused regions on any pieces of the stem which transform into earthy coloured patches that are noticeable from distance. At first, the bit of the stem above or beneath the patches looks solid. At last, the contamination supports the stem totally which reaches out as much as a foot or more. The decay makes the tissues become delicate and effectively strip off into shreds. The bits over the influenced part may eventually wither, die back and break away (Tripathi *et al.*, 2013).

The outside of the influenced parts is covered with white stands of mycelia which structure pad in the axils of the branches. Dark-hued sclerotia might be seen on this mycelial mat. The substance district might be loading up with this hard scleroria. The sclerotia were additionally seen in bolls. The sickness begins in the long stretch of December and spread with time till the boll arrangement.

Causal organism

Stem rot is caused by the fungus *Sclerotinia sclerotiorum*

Disease cycle

As there are no secondary inoculums produced by *Sclerotinia sclerotiorum*, it can be described as monocyclic. The fungus produces sclerotium or sclerotia either on or inside the plant tissue. The sclerotia get by in the dirt and it sprouts and created apothecia in which asci and ascospores are framed. Upon discharge the ascospore cause contamination.

Epidemiology

The disease shows up in serious structure during the cooler long stretches of December and January. The Relative humidity for December in 2012 and January in 2013 was 76% and 72% separately, with related mean month-to-month temperatures of 18.0^0C and 17.0^0C. The temperature ranges from 15-20^{0}C under moist conditions are suitable for the development of the fungus.

Management

Stem rot in mesta can be managed by cultural, biological and chemical practices. Cultural practices include crop rotation, planting crops at lower densities and higher row spacing to create microclimates that are less favourable for disease development. Tillage operation can reduce the number of pathogen spores.

The use of systemic and contact fungicides can reduce the development of stem rot disease. Biological agents are also used for controlling the disease.

3. Mesta leaf blight (*Phyllosticta hibiscini*)

The infection was first detailed in Nigeria. In India, the occurrence of leaf scourge of roselle (*Hibiscus sabdariffa* L.) brought about by has been recorded first time from Barrackpore, West Bengal, India during September 2014.

Symptoms

The infection begins as a stained water-doused region generally from the edge of the leaf which expanded towards the internal bearing and contaminates the petiole, through which it moves towards the stem. The infected leaf and petiole began yellowing lastly tumbling off. The contamination spread exceptionally quick under high moisture and precipitation condition and plants become defoliated. Under the dry condition, the diseases become confined and dim/dark shaded spots like pycnidia are created on the leaf.

Causal organism

This disease is caused by *Phyllosticta hibiscini*.

The sickness is brought about by *Phyllosticta hibiscini*, dirt-borne microbes. The microbe produces a white-shaded province with distinguished ring-like zones on which dark-hued speck-like pycnidia are created inside 72h at 28±1°C. A tiny investigation uncovered plentiful hyaline profoundly stretched (for the most part dichotomous) mycelia of 4-6μ. Various dim/dark coloured pycnidia were taken note in the PDA media that delivered hyaline circular, for the most part, single-celled conidia estimating around 8-12 μ, round about thick-walled chlamydospores were moreover noticed under the magnifying instrument.

Epidemiology

The infection is supported by high mugginess (75-95% RH), high temperature (30-370C) and high precipitation. Under dry conditions the disease becomes limited. With the increment in the crop, age defoliation happens from the foundation of the plant making the plant debilitated by all accounts.

Management

A safe assortment like AMV 3 is to be developed. Seed treatment ought to be finished with thiram @ 1.25 mg/kg of seed. A prophylactic shower with Copper oxychloride at 0.075% may likewise be finished.

4. Yellow vein mosaic virus

Yellow vein mosaic virus is a serious disease in India in mesta crops. This disease is endemic in a different part of India. Due to this disease in the mesta crop, it reduces the fibre quality and seed yield and thus becomes a warning to the production.

Symptoms

The symptoms of this disease are the yellowing of veins and veinlets. Complete chlorosis is observed at the affected leaves with the advancement of infection. The plant does not even flower if the virus occurs at its early stages. The diseased plant remains stunted and reduced plant size. Yellow vein mosaic virus is caused by begomovirus. This disease is transmitted by whitefly (*Bemisia tabaci*) under natural conditions. YVM virus can be detected by electron microscopy, molecular techniques and southern hybridization (Roy *et al.*, 2009)

Management

To reduce the yellow vein mosaic virus, the vector population should be controlled. The affected plant should be uprooted and burned as and when infections are observed in the plant.

5. Root-knot nematode

Root-knot nematode is an important and economically important nematode pest in mesta. It causes damage to the plant up to 15-20%.

Symptoms

Root-knot does not produce any specific above-ground symptoms. Affected plants have an unthrifty appearance and often show symptoms of stunting, chlorosis or wilting. Symptoms are severe when plants are infected soon after planting.

Causal organism

Root-knot nematode is caused by *Meloidogyne incognita.*

Meloidogyne incognita eases the entry of other fungal and bacterial pathogens and cause heavy damage by creating injury to the roots.

Management

The root-knot nematode can be controlled by cultural practices like thinning, weeding, and removal of stubbles. It can be also controlled by some biocontrol agents. Besides these major diseases, there are some minor diseases of mesta.

6. Anthracnose

Anthracnose mainly occurs in cannabinus mesta. The terminal bud is infected initially. Necrotic spots are developed in young leaves. Black lesions appear on stem infections which later form cavities.

7. Seedling rot

This disease is mainly susceptible to cannabinus mesta. In this disease black lesions is appeared initially on cotyledons which rot and wither.

Seedling rot can be controlled by applying potassium fertilizer and maintaining proper drainage. The acidity of the soil should be checked.

8. Collar rot

This disease is caused by both the species of the mesta, i.e. *cannabinus and sabdariffa*. When this disease the plant, a black seated lesion appears on the ground level of the stem.

For the management of this disease, there is no control measures have been developed. However, waterlogging should be avoided to prevent this disease.

Conclusion

Mesta is a bast fibre and is used as a substitute for jute. Its production is less due to abiotic and biotic stresses. The fungal disease mainly foots and stem rot and the viral disease yellow vein mosaic virus causes many infections to this crop. Due to these diseases, the crop production decreases.

Biotic stresses can be controlled through biological, chemical and cultural methods. Yield should be achieved by the development of resistant varieties. Biocontrol agents should be used to control the disease instead of chemical control.

References

Anonymous. (2014-15). Central Research Institute for Jute and Allied Fibres, Barrackpore, Kolkata. pp. 47

Chatterjee, A. and Ghosh, S. K. (2008). Alterations in biochemical components in mesta plants infected with yellow vein mosaic disease. *Brazilian journal of plant physiology*, **20**, 267-275.

De, R.K. and Mandal, R.K. (2007a). Effect of seed treatment with fungicides on foot and stem rot disease caused by *Phytophthora parasitica var. sabdariffae* in *Hibiscus sabdariffa. Journal of Interacademicia.* **11**:161-165

De, R.K. and Mandal, R.K. (2007b). Epidemiology of foot and stem rot and collar rot diseases of HS and HC mesta. In: National symposium on microbial diversity and plant health. Bidhan Chandra Krishi Viswavidyalaya, Mohanpur. pp.30

http://www.theplantlist.org/tpl1.1/record/kew-2849657

https://en.m.wikipedia.org/wiki/Kenaf

https://vikaspedia.in/.../package-of-practices/commercial-crops/mesta

https://www.apnikheti.com/en/pn/agriculture/crops/greenmanure/mesta#
Typesofvarieties

https://www.slideshare.net/harryraj/mesta-fiber-crop

Islam, M. S., Uzzal, M. S. I., Mallick, K. and Monjil, M. S. (2013). Management of seed mycoflora of mesta (*Hibiscus sabdariffa*) by seed washing, garlic extract and knowin. *Progressive Agriculture, 24*(1-2): 1-7.

Mahapatra, B. S., Mitra, S., Ramasubramanian, T. and Sinha, M. K. (2009). Research on jute (*Corchorus olitorius* and *C. capsularis*) and kenaf (*Hibiscus cannabinus* and *H. sabdariffa*) present status and future perspective. *Indian Journal of Agricultural Sciences, 79*(12): 951-967.

Roy, A., Acharyya, S., Das, S., Ghosh, R., Paul, S., Srivastava, R. K. and Ghosh, S. K. (2009). Distribution, epidemiology and molecular variability of the begomovirus complexes associated with yellow vein mosaic disease of mesta in India. *Virus Research*, **141**(2), 237-246.

Tripathi, A.N., Sarkar, S.K., Sharma, H.K. and Karmakar, P.G. (2013). Stem rot of roselle: A major Limitation for Seed production. *Jaf News*. 11, 14.

Chapter - 11

Diseases of Sunn hemp (*Crotalaria juncea* L.) and their Integrated Management

Lipa Deb, Pranab Dutta and Jyotim Gogoi

School of Crop Protection, College of Post Graduate Studies in Agricultural Sciences, Central Agricultural University, Umiam, Meghalaya-793103

Introduction

Sunn hemp (*Crotalaria juncea* L., Family: Fabaceae, subfamily: Faboidae) originated in India (Montgomery, 1954) and has been known as brown hemp, Indian hemp, and madras hemp in tropical Asian countries. Etymologically, the genus named "*Crotalaria*" meaning "rattle" due to the rattling noise made by seeds, when shaken inside mature pods, whereas, the species named "*juncea*" by Carolus Linnaeus due to its resemblances with Spanish broom, *Spartium junceaum* L. from Mediterranean region (Kundu, 1964; White and Haun, 1965). Botanically, Sunn hemp is cross-pollinated, short day, annual crop with 1 to 4 m erect shrub, cylindrical and ribbed stems, simple, elliptic to oblong-lanceolate shaped, spirally arranged leaves with a long taproot and well-developed lateral roots. The crop bears terminal, open, raceme inflorescence with deep yellow, indeterminate, photoperiod-sensitive flowers up to 25 cm in length producing small, flattened, kidney-shaped seeds.

An economical perspective, the major production of sunn hemp in tropical and sub-tropical regions revolves around an annual, renewable, multipurpose fiber crop due to high lignin content, where the stem is used for bast and woody core fibers at the proportion of 15-20% dry weight of bark in the total stalk (Cunningham *et al.*,

1978). The bast fiber obtained is used for manufacturing a wide variety of products *viz.*, tissue paper, marine cordage, fishing net, bank currency, carpets, paper, string, marine cordage, belting, rope, webbing hose pipe, canvas etc. Due to the higher yield of bleachable sulfate pulps, pulp strength properties and higher bast fiber length to width ratio, sunn hemp possess excellent applications in paper making and biofuel industries. In countries like India, Bangladesh, Pakistan and Brazil, sunn hemp is grown as fiber, fodder and green manure crop in light loam, moderately rich well-drained soil, having an average temperature of 23-29.4 °C, high humidity and rainfall of 170-200 mm during Kharif season (Dempsey, 1975). Additionally, sunn hemp is also grown as a promising source of green manure crop, when ploughed during the early flowering stage i.e., 1.5-2.5 months old resulted in the highest nitrogen recovery of 134-147 lb/acre of nitrogen and 3 tons/acre of dry organic matter at 60 days of growth (Shil *et al.*, 2018). Being a tropical legume crop, sun hemp shows the ability to undergo symbiosis with rhizobacteria *viz.*, *Rhizobium, Bradyrhizobium, Azorhizobium, Bradyrhizobium* etc. eliciting lobed root nodulation up to 2.5 cm length and fix 60% of the total biological nitrogen per annum (Herridge *et al.*, 2008).

Besides, in vegetables and field cropping systems, sunn hemp is used as a natural suppressor of plant parasitic nematodes e.g., root-knot nematode *Meloidogyne incognita* by enhancing natural enemies of PPNs that trap nematodes or feed on their eggs resulting in few numbers of galls (Rotar and Joy, 1983). They are also incorporated in fields to suppress weeds and slow down soil erosion (Wang and Mcsorley, 2009). Wang *et al.* (2002) reported the use of sunn hemp against several sedentary endoparasitic nematodes such as soybean cyst (*Heterodera glycines*), reniform nematodes (*Rotylenchulus reniformis*), as well as migratory nematodes such as stubby root (*Paratrichodorus minor*), sting *(Belonolaimus longicaudatus)*, burrowing *(Radopholus similis)*, dagger *(Xiphinema americanum)* nematodes. In India, sunn hemp production is subjected to a variety of diseases caused by various microorganisms' *viz.* fungi, viruses and phytoplasma. The present review gives a brief highlight of about most important sunn hemp diseases *viz.* vascular wilt, anthracnose, mosaic, leaf curl, phyllody *etc* along with their characteristic symptoms, causal organisms, disease cycle, epidemiology as well as integrated management approaches.

Vascular wilt of Sunn hemp, *Fusarium udum*

Economic importance

Fusarium wilt of sunn hemp have been reported from different countries *viz.*, Taiwan (Wang and Dai, 2018) and Korea (Choi *et al.*, 2018). However, in India, the disease incidence of 10-12 % has been reported from major sunn hemp growing states such as Bihar, Uttar Pradesh, Madhya Pradesh and Maharashtra, which can increase up to as high as 60-80% under favourable conditions.

Symptoms

Initially, the affected plants showed general yellowing starting with the lower leaves of the plant, which gradually wither and wilt by drooping at tips, hanging down and turning brown. Within a day or two, wilted plants defoliate and eventually die. However, internal tissues of affected plants such as the main tap and lateral roots showed gradual discolouration of vascular bundles (Sarkar and Gawande, 2016). At the early stage of the fungus, tissue discolouration starts from the tip portion of lateral roots, subsequently reaching the crown region and continuing upward from one side of the stem attacking vascular bundles of the meristem. At a later stage, sporodochium of fungi with white to pinkish fungal spore masses and dark perithecia are formed on a dead portion of the stem. At matured crop stage, the fungus also affected pods and infected discoloured seeds may be also noticed in the field.

Causal organism

Taxonomic position of wilt fungus of sunn hemp:

     Domain: Eukarya

     Kingdom: Fungi

     Phylum: Ascomycota

     Sub Phylum: Pezizomycotina

     Class: Sordariomycetes

     Subclass: Hypocreomycetidae

     Order: Hypocreales

     Family: Nectriaceae

     Genus: *Gibberella*

     Species: *G. indica*

     Anamorph: *Fusarium udum*

Fusarium udum E.J. Butler f. sp. *crotalariae* (G.S. Kulkarni) Subramanian (Mitra, 1934; Kundu, 1964; Purseglove, 1968) having teleomorph: *Gibberella indica*, is a soil-borne fungus characterized by hyaline, slender, branched, aerially growing hyphae producing three types of spores *viz.*, chlamydospores, microconidia and macroconidia. The macroconidia are hyaline, falcate-shaped with indistinct foot cells, usually, 1-4 septate, curved to hooked apical cells measuring 21-41 x 3-4.5 µm in size are borne on short conidiophores that detached after abscission. While, microconidia are oval- to reniform-shaped, elliptical, hyaline singly but salmon-pink coloured in mass, are borne on false heads by monophialides conidiogenous cells on single or clustered, vertically branched conidiophores, usually aseptate, measuring 6-10 x 2-4 µm in size (Choi *et al.*, 2018; Wang and Dai, 2018). Thirdly, chlamydospores of oval or spherical shape are produced abundantly in hyphae, either singly or in chains, usually thick-walled, measuring 5-10 µm in size (Holliday, 1980).

Disease cycle

Fungal propagule overwinters in the soil and crop residues as facultative parasites, which in presence of host cues, produce enormous spores *viz.*, macro-, microconidia and mycelia in sporodochium. The pathogen attack plant either through thinner root and rootlets or through cracking at the basal portion of stems, thereby, occurring inter and intracellularly and ecto-trophically on the collar region and roots of infected plants (Choi *et al.*, 2018). The fungus also produces perithecia in the crown region of the infected plants, whereas, conidia and chlamydospores served as resting spores for the long-term survival of the fungus. In infected plants, saprobic mycelium produces conidia that germinate into germ tubes, penetrate roots and colonizes xylem vessels, thereby, causing wilt symptoms. Transmission of the pathogen is facilitated by farm implements, irrigation water, infected pods and discoloured seeds that may disseminate and initiate secondary infection in the field.

Epidemiology

The incidence of vascular wilt of sunn hemp is influenced by soil temperature and retentive nature of the soil, whereas, relative humidity and water content were found non-significant in wilt disease development. Decrease in soil temperature (between 20-30 ºC), sandy soil, high rainfall and increasing plant maturity are the most conducive factors for the spread of disease favouring higher wilt disease incidence in low-lying areas. Being a soil-borne pathogen, the fungus spreads for about 3 m through the soil, surviving up to a depth of 120 cm for about 3 years on infected plant debris (Naik, 1993). The disease incidence was recorded reportedly lower in early sown crops during mid-April. In a study at Sunn hemp Research Station, Pratapgarh, Uttar Pradesh, wilt diseases were observed starting from 62 days after sowing (DAS) that increased significantly with an increase in the age of the crop.

F. udum affects many species of leguminous plants *viz.*, *Vigna angularis*, *Phaseolus vulgaris*, *Pisum sativum*, *Sesbania cannabina*, *Glycine max* and *Cajanus cajan* (Wang and Dai, 2018).

Anthracnose of Sunn hemp, *Colletotrichum curvatum*

Economic importance

In India, the anthracnose disease of Sunn hemp was reported from all major sunn hemp growing states such as Pusa, Bihar with severity as high as 60 -80%.

Symptoms

Anthracnose of sunn hemp is a foliar disease that initially appears on cotyledon of seedlings as soft-discoloured areas, from which it subsequently infects, forms brown spots and spreads towards the stems and other growing points of plants except for underground parts. Plants are susceptible at all growing stages but, symptoms most frequently occur on stems, leaves and pods. In young plants, affected seedlings drop from the petiole below the cotyledon, and ultimately dies within two days (Sarkar and Gawande, 2016). Whereas, in older plants, infection is restricted on stems and leaves, having greyish brown to dark brown spots of round to irregular shape that initially occur on one side of leaves, later enlarging and extending to the opposite side. The fungal infection further spreads downwards, where acervulli are formed on infected areas within two days. In severe conditions, spots coalesce covering the entire leaf surface, turn brown and infected leaves fall off.

Causal organism

Taxonomic position of anthracnose fungus of sunn hemp:

> Domain: Eukarya
>
> Kingdom: Fungi
>
> Phylum: Ascomycota
>
> Class: Sordariomycetes
>
> Order: Glomerellales
>
> Family: Glomerellaceae
>
> Genus: *Colletotrichum*
>
> Species: *C. coccodes, C. atramentarium, C. dematium, C. curvatum, C. crotolariae*

Several species described on *Crotalaria* sp., *C. curvatum* Briant and Martyn and *C. crotalariae-junceae* Swada were reported synonym with *C. truncatum* having 18-26 x 3-4.8 μm size conidia with ling parallel walls rising shortly after base. Whereas, *C. crotalariae* Petch reported on *Crotalaria striata* showed straight conidia, and are regarded as synonyms of *C. coccodes*. In other species, conidia of *C. gangeticum* Pavgi and U.P. Singh reported on *Crotalaria* spp. in India, showed small (14.3-17.1 x 2.8-4.3 μm size), pointed ends with dark-brown setae broadened at the tip, in contrast, *C. verruculosum* bears conidia with rounded ends and brown setae, thin at the apex. *C. dematium* is characterized by crowded, black, oval to elongate, hemispheric to truncate-conical, erumpent acervulli on stomata bearing numerous needle-like setae measuring 60-300 x 3-8 μm in size. Conidia are bluntly tapered, curved, unicellular, hyaline, measuring 17-31 x 3-4.5 μm are born singly on conidiophores.

Disease cycle

Pathogen perpetuates in the form of mycelium in infected crop residues or seeds that act as primary inoculum and may cause pre and post-emergence damping-off of seedlings. In young actively growing plants, fungal mycelia remain localized in the tissues of lesions at the site of initial infection, that later become systemic, spreading both inter- and intracellularly through stems, vascular bundles, petioles, leaves, pods and seeds. Numerous single-celled hyaline conidia formed on acervulli in the epidermis of the diseased area further lead to the secondary spread of the infection.

Epidemiology

Disease development and perpetuation of anthracnose are favoured by cloudy weather, continuous rain and thickly populated crops. Fungal propagule i.e. conidia germinate and form appressoria at a temperature below 35 °C, accompanied by the wet surface of the plant or relative humidity above 70.00%. However, transmission and spread of fungal spores to adjacent plants are facilitated by rain splashing. Secondary stem and pod infection are favoured by humid, warm and moist weather prevailing during the later stage of pod maturation (Bhale *et al.*, 1998; Neergaard *et al.*, 1999).

Sunn hemp mosaic and southern sunn hemp mosaic, *Sunn hemp mosaic virus*

Economic importance

Sunn hemp mosaic and southern sunn hemp mosaic disease have been reported as major destructive diseases from major sunn hemp growing states, and are of great economic importance (Capoor, 1962).

Symptoms

In young plants, the first visible symptoms of sunn hump mosaic disease appear as mottling of leaves within 10-12 days after inoculation. At a later stage, the disease progress into very prominent patches of light and dark areas. Infected plants remain shorter in height, and bear smaller leaves with abnormal lamina than normal, producing dark green areas on the upper leaf surface with depressions on the lower surface, ultimately reducing seed and fiber yields. Whereas, symptoms of southern sunn hemp mosaic disease *viz.*, faint, discoloured patches on young leaves, appear initially within 9-20 days of a favourable condition. At later stages, mosaic symptoms develop into distinct patterns with intense puckering and blistering of leaves that is subsequently accompanied by thin elongated enations running parallel from the underside of the leaves. At advanced stages of crop growth, the characteristics of mosaic symptom, along with mottle and leaf distortion takes place at varying degrees, therefore, making infected plants dwarf-bearing dwarf leaves and scanty flush of flower resulting in poor pod setting and seed yield.

Causal organism

Taxonomic position of sunn hemp mosaic virus:

Domain: Virus

Order: *Martellivirales*

Family: *Virgoviridae*

Genus: *Tobamovirus*

Species: *Sunn hemp mosaic virus*

Sunn hemp mosaic virus (SHMV) is a positive-sense single-stranded RNA virus, spherical particles, measuring 26-40 µm in size. The virus strain has been found morphologically similar to the cowpea mosaic virus. While, southern sunn hemp mosaic virus (SSMV) are rod-shaped particles, measuring 300 µm x 18 µm in size with short virion of 32-34 nm, 6.3-6.6 kb genome and 17-18 kDa single structural protein (Plate 5). The virus particle was reported to be highly stable with DEP 10-7, TIP-950C and ageing under *in vitro* conditions for 6 years and can withstand complete desiccation.

Disease cycle and epidemiology

Sunn hemp mosaic virus is reported to be mechanically transmitted from infected to healthy plants, among various hosts such as *Nicotinia tabacum, Pisum sativum, Lycopersicon esculenta, C. mucronate, Datura strumonium* and *Cyamopsis tetragonoloba*. From earlier studies conducted at the Sunn hemp research station and farmer's field of Pratapgarh, U.P., the severity of the sunn hemp mosaic virus correlated with the onset of monsoon (Sarkar and Gawande, 2016). Being a sap-transmitted virus with a wider host range, the virus was also reported to be harboured in wild or weed hosts. In addition, the higher or optimum dose of nitrogen and phosphorus was reported to increase the concentration and spread of the virus within the plant, whereas, the application of potassium further reduces the concentration of the virus and spread of the disease.

The incidence of southern sunn hemp mosaic (SSMV) disease was reported to be higher during monsoon months with temperatures ranging between 22-41 °C and relative humidity of 80.00%. Transmission of the virus was reported to be mechanical as rubbing of leaves without any abrasion, wedge or patch grafting with diseased grafts but not through seed, soil and vectors. Various hosts for the virus include *Phaseolus vulgaris, Solanum nigrum, Datura strumonium, C. retusa, C. spectabilis, V. unguiculata, Cajanus cajan, N. glutinosa, N. tabacum* etc. In an earlier study by Solomon (1972), reported manual inoculation of SSMV in sunn hemp seedlings resulted in invariably 100.00% infection caused by maximum virus titre at 10 days after inoculation, which remained higher up to the emergence of inflorescence primordia. However, translocation studies and movement patterns of SSMV by Solomon and Sulochana (1979) further reported systemic spread of the virus in emerging stems, leaves and hypocotyls indicating movement of virus particles from the site of inoculation i.e., primary leaves lamina through petioles, upward and downward movement in stems as well as virus multiplication within 24 h of inoculation.

Sunn hemp phyllody, *Candidatus* Phytoplasma phoenicium-related phytoplasma

Economic importance

In India, sunn hemp phyllody has been first reported from the Central Research Institute of Jute and Allied Fibers (CRIJAF) research farm, Barrackpore in 2016 by Biswas *et al.* (2018). Later, sporadic incidences of sunn hemp phyllody disease have been reported from major sunn hemp growing states of India. The disease has also resulted in a considerable loss in seed yield up to 2-4% by affecting seed crop (Sarkar, 2010).

Symptoms

Typical symptoms of sunn hemp phyllody include yellowing of apical leaf portions at the initial stage which is followed by the formation of a big bud at the terminal raceme of the infected plant. Later, entire meristem florets get converted into vegetative stage leading to the formation of the dwarf shoot by shoot proliferation with reduced size leaves, and chlorotic leaves, giving phyllod, witches broom and little leaf appearance of bud in the infected plant (Win *et al.*, 2011). The disease is also characterized by bunchy top symptoms, bending, discolouration of leaf veins and flattened stem appearance causing stem fascination in sunn hemp plants.

Causal organism

Taxonomic position of anthracnose fungus of sunn hemp:

     Domain: Bacteria

     Phylum: Firmcutes

     Class: Mollicutes

     Order: Acholeplasmatales

     Family: Acholeplasmataceae

     Genus: Phytoplasma

     Species: *Candidatus* Phytoplasma Phoenicium

The causal organism for sunn hemp phyllody disease was identified as *Candidatus* Phytoplasma phoenicium (CaPphoe) which showed similarity with other phytoplasma strains clustered under 16 SrIX-E subgroups i.e., "<i>Ca.</i>" (Biswas *et al.*, 2018). They are characterized as pleomorphic, wall-less fastidious vascular bacteria (formerly mycoplasma-like organisms, MLOs), with small genome sizes ranging from 530-1350 kilobases (Marcone, 2014), exist as obligate plant pathogens exclusively living inside phloem sieve tubes.

Disease cycle and epidemiology

Transmission of phytoplasma causing sunn hemp phyllody was facilitated by sap-sucking insect vector Indian cotton jassid, *Amrasca biguttula biguttula* (Lee *et al.*, 2000). Adult cotton jassid (Hemiptera: Cicadellidae) is a long, slender (2.6 mm), yellowish-green leaf hopper, having a pale-green head, membranous transparent wings, with conspicuous black dots on either side of the head and tip of the forewing. In Sunn hemp research station, Pratapgarh, Uttar Pradesh, Sarkar (2000) reported similar disease-causing witches broom transmitted mechanically, occurring usually after application of tetracycline @ 500 ppm after 30-60 days after sowing. Phytoplasma under taxonomic group 16SrIX has been reported to be associated with diseases of several crops in wide geographical areas such as almond, pigeon pea, peach, nectarine orchards, plum, cherry, apricot, grapevine *etc* (Bertaccini *et al.*, 2014).

Sunn hemp leaf curl

Economic importance

Sunn hemp leaf curl is one of the most devastating diseases in sunn hemp growing regions limiting crop cultivation, especially during monsoon season (Khan *et al.*, 2002; Raj *et al.*, 2003). Natural incidence of the disease has been reported at about 75.00% in different districts in eastern Uttar Pradesh (Sarkar *et al.*, 2015), causing fiber loss of up to 90.00% from Sunn hemp research station, U.P.

Symptoms

Sunn hemp leaf curl disease initiates as light faint mosaic mottling and leaf curling approximately after 15 days of germination. At a later stage, the symptom intensifies into severe mosaic, yellow mottling, and upward and downward leaf curling resulting in reduction in the plant height and leaf size of the affected plant (Kumar *et al.*, 2010). The disease is also characterized by interveinal chlorosis of leaves, yellowing and shortening of leaves (Khan *et al.*, 2002).

Causal organism

Several studies have reported the association of geminivirus *i.e.* Begomovirus with sunn hemp leaf curl disease (Brunt *et al.*, 1996). Taxonomic position of sunn hemp leaf curl virus:

Domain: Virus

Order: Geplafuvirales

Family: *Geminiviridae*

Genus: *Begomovirus*

Begomoviruses consist of spherical bipartite particles and ssDNA as a genome with two similar-sized DNA components (DNA A and DNA B). The DNA A component encodes for a replication-associated protein (Rep), replication enhancer protein (REn), coat protein (CP) and transcription activator protein that are essential for viral DNA replication and gene expression. While, the DNA B component encodes nuclear shuttle protein (NSP) and movement protein (MP) essential for systemic infection of plants (Gafni and Epel, 2002).

Epidemiology

Begomovirus is transmitted from diseased to healthy plants by whiteflies, *Bemisia tabaci* (Hemiptera: Aleyrodidae) facilitating the secondary spread of the disease. It is a polyphagous insect distinguished into 1 mm long adult, male slightly smaller than females and entire body and wings are covered with white to slightly yellowish powdery waxy secretion.

Integrated disease management approach in sun hemp

Cultural approach

» Very strict quarantine measures should be practised to prevent the entry of pathogens in the form of infected planting materials into disease-free areas.

» Selection of the proper site with adequate drainage and preventive phytosanitary measures such as cleanliness, sound hygiene, soil fumigation and treated irrigation water should be practised.

» Use clean and healthy seeds to avoid seed-borne infections.

» Removal of plant debris, stubbles, and weeds that serve as the primary source of inoculum. Soil should be prepared by deep ploughing and exposure to the sun to destroy soil-borne inoculum.

» Application of lime @ 2-4 tonnes/ha a month before sowing is preferred for obtaining neutral soil pH in the range of 6.5-7.5 as acidic soil is one of the perpetuation factors for many diseases.

» Avoid indiscriminate use of nitrogenous and phosphate-based fertilizers. Excess application of nitrogen fertilizers though induces vegetative growth but makes plants susceptible to diseases at a later stage.

» In North Indian conditions, planting of sunn hemp from April to August has been recommended for the escape of disease, while, delay in sowing can lead to an increase in susceptibility to wilt disease.

» Early sowing during mid-April to mid-May also avoids predisposing factors of disease *viz.*, high rainfall and high humidity favouring disease incidence in sunn hemp.

» Avoid close spacing and dense cropping, as it is more conducive to disease. Adequate spacing of 30 cm row to row and 5-6 cm plant to plant in line sowing at the time of thinning is recommended.

» Intercropping of sunn hemp with sesame and sorghum along with the additional application of zinc has been reported to reduce wilt disease incidence of sunn hemp.

» Monitoring of field and seed crops should be done continuously at regular intervals to identify the infection at the earliest possible.

» Any suspected diseased or infected plant materials should be removed and destroyed burning can further reduce the spread of disease.

» Avoid water logging conditions, as a basal region of sunn hemp produces adventitious roots that impair fiber quality.

Chemical approach

» Application of Carbendazim or Thiram as seed treatment (@ 2g/Kg) followed by foliar spray with Carbendazim 50 WP (@ 2g/l) at 15 days interval has been recommended against fungal diseases of Sunn hemp.

» In addition, seed treatment and soil application with zinc sulphate ($ZnSO_4$) with neem cake also confer protection against the wilt of sunn hemp.

» Eradication of diseased portion *viz.* infected pods, leaves or twigs, stems should be done by cutting with a sharp knife/ chisel and cleaning the wounds with 0.1% mercuric chloride or 1% potassium permanganate solution followed by application of Bordeaux paste or ridomil paint.

» Effective management of vector of sunn hemp phyllody *Bemisia tabaci* should be done by application of Thiamethoxam 25 WG @ 0.25% and Imidacloprid 17.8 SL @ 0.3%.

Biological approach

» Application of *Trichoderma viride* as seed treatment and soil drenching two times i.e., at 15 and 30 DAS can induce a primed state in sunn hemp against various diseases.

» Other biocontrol agents such as *Aspergillus niger* and plant growth promoting rhizobacteria (PGPR) were also reported in effective management of seed, soil and foliar diseases of sunn hemp.

Resistant varieties

» Considerable efforts on screening of germplasm from wild *Crotalaria* species *viz., C. mucronate, C. striata, C. brevidence, C. verucosa* etc. conferring resistance against various biotic factors in sunn hemp have been made.

» Use of K-12 yellow, a resistant variety against Sunn hemp vascular wilt have been recommended from a selection of K-12 (black) variety.

» Several varieties *viz.,* SH-4, SUN 053, JRJ 610 and SUIN 037 have been recommended and released as tolerant against wilt and mosaic disease of sunn hemp under natural conditions.

Future approaches

» The management of various sunn hemp diseases is usually subjected to repeat fungicide application resulting in higher cost and poisonous residual effect in the environment, therefore, new strategies for the development of durable control is a must.

» Biological control is an important alternative to chemical fungicides and the search for new broad-spectrum antagonists that are effective against multiple pathogens is essential aiming for organic agriculture and long-term sustainable goals.

» Efficient local strains of antagonistic microbes should be identified, and evaluated for their efficacy in disease control, mass multiplied with an aim for enhancement in productivity and quality of fibres.

» Recent works on the isolation of endophytes from sunn hemp have provided a new dimension in crop protection by enhancing inbuild innate resistance in crops.

» Knowledge of biology and variability of pathogens is a prerequisite for breeding programs aimed at obtaining durable resistance.

» Studies on the ecology of pathogens and their epidemiological aspects as well as proper prediction, weather-based models and forecasting systems are needed to be developed. This will prevent the disease outbreak in advance, and lessen the impact on crop yield and fibre quality in different agro-climatic conditions.

»   Systematics and etiological studies should be given more priority to learn about the life cycle of the pathogen to figure out accurate and specific curative measures.

»   Pathogen-vector interaction and potential helper factors should be identified and characterized, which will provide new insights on epidemiology and new control strategies.

»   Effect of climate change on dynamics of plant pathogens, pests and antagonistic microorganisms should be given emphasis in order to study their impact on biological and behavioural activity.

»   An expert system of Sunn hemp as an advisory for disease management by analysing visuals-, data-, and short and long-term crop protection measures should be developed and adopted.

Conclusion

Sunn hemp in India is severely infected by various diseases such as wilt by *Fusarium* spp., anthracnose by *Colletotrichum* spp. and several viral and phytoplasma diseases such as mosaic, leaf curl as well as phyllody. Every year, such entities lead to huge losses in sunn hemp production nationally and globally refraining the fiber industry worldwide. Management of these diseases is essential to provide increased and sustainable fiber production throughout India and the world. Understanding characteristic detection symptoms of these diseases at all the stages of plant growth can help to control diseases way earlier than their potential infectious stage. Knowledge about causal organisms, their life cycle, and predisposition factors for disease incidence as well as disease cycle aids in framing stringent diagnostic procedures and forecasting its epidemiological outbreak well in advance. Intensive focus on integrated disease management (IDM) combining the use of disease-free planting materials, utilization of bio-agents, improved cultural practices, development of resistant varieties and chemical control strategies in a holistic way rather than a single component strategy proved to be more effective and sustainable for disease management.

References

Bertaccini, A., Duduk, B., Paltrinieri, S. and Contaldo, N. (2014). Phytoplasmas and phytoplasma diseases: a severe threat to agriculture. *American J. Plant Sci.*, 5(12): 1763-1788.

Bhale, M.S., Bhale, U. and Khare, M.N. (1998). Diseases of important oilseed crops and their management. *In:* Khurana, S.M.P. (ed.) Pathological problems of economic crop plants and their management. Jodhpur, Scientific Publishers (India). Pp. 251-279.

Biswas, C., Dey, P., Meena, P.N., Satpathy, S. and Sarkar, S.K. (2018). First report of a subgroup 16SrIX-E (*Candidatus* Phytoplasma phoenicium-related phytoplasma associated with phyllody and stem fasciation of sunn hemp *Crotalaria juncea* L. *Plant Dis.*, doi.org/10.1094/PDIS-04-17-0584-PDN.

Brunt, A.A., Crabtree, K., Dallwitz, M.J., Gibbs, A.J. and Watson, L. (1996). *Viruses of Plants. Descriptions and Lists from the VIDE Database.* Wallingford, UK, CAB International.

Capoor, S.P. (1962). The movement of tobacco mosaic virus and potato virus x through tomato plants. *Ann. Appl. Biol.*, 36: 307-319.

Choi, H.W., Hong, S.J., Hong, S.K., Lee, Y.K. *et al.* (2018). Characterization of *Fusarium udum* causing *Fusarium* wilt of sunn hemp in Korea. *Kor. J. Mycol.*, 46(1): 58-68.

Cunningham, R.L., Clark, T.F. and Bagby, M.O. (1978). *Crotalaria juncea*-annual source of papermaking fiber. *TAPPI*, 61: 37-39.

Damm, U., Woudenberg, J.H.C., Cannon, P.F. and Crous, P.W. (2009). *Colletotrichum* species with curved conidia from herbaceous hosts. *Fungal Div.*, 39: 45-87.

Dempsey, J.M. (1975). Fiber Crops. The University Presses of Florida, Gainesville, Florida.

Gafni, Y. and Epel, B.L. (2002). The role of host and viral proteins in intra- and inter- cellular trafficking of geminiviruses. *Physiological and Molecular Plant Pathology*, 60(5): 231-241.

Herridge, D.F., Peoples, M.B. and Boddey, R.M. (2008). Global inputs of biological nitrogen fixation in agricultural systems. *Plant Soil*, 311: 1–18.

Holliday, P. (1980). Fungus diseases of tropical crops. Fungus diseases of tropical crops. Cambridge, UK: Cambridge University Press.

Khan, J.A., Siddiqui, M.K. and Singh, B.P. (2002). The natural occurrence of a begomovirus in sunn hemp *(Crotalaria juncea)* in India. *Plant Pathol.*, 51: 398.

Kumar A., Kumar, J., Khan, Z.A., Yadav, N., Sinha, V., Bhatnagar, D. and Khan, J.A. 2010. Study of beta satellite molecule from leaf curl disease of sunn hemp (*Crotalaria juncea*) in India. *Virus Genes*, 41:432-40

Kundu, B.C. (1964). Sunn-hemp in India. Proc. Soil Crop Soc. Florida. 24: 396-404.

Lee, I.M., Davis, R.E. and Gundersen-Rindal, D.E. (2000). Phytoplasma: phytopathogenic mollicutes. *Ann. Rev. Microbiol.*, 54: 221-255.

Marcone, C. (2014). Molecular biology and pathogenicity of phytoplasmas. *Ann. Appl. Biol.*, 165(2): 199-221.

Montgomery, B. (1954). Sunn fiber. In: Mauersberger, H.R. (ed.) Mathew's textile fibres. 6th ed. Wiley, New York. Pp. 323-327.

Naik, M.K. (1993). Ecology and integrated disease management of fusarium wilt of pigeonpea. Legumes Pathology Progress Report No. 19. Patancheru, India: ICRISAT.

Neergaard, E. De., Tornoe, C. and Norskov, A.M. (1999). *Colletotrichum truncatum* in soybean: studies of seed infection. *Seed Science and Technology*, 27: 911-921.

Raj, S.K., Singh, R., Pandey, S.K. and Singh, B.P. (2003). Association of a geminivirus with a leaf curl disease of sunn hemp *(Crotalaria juncea)* in India. *Eur. J. Plant Pathol.*, 109: 467-470.

Rotar, P.P. and Joy, R.J. (1983). 'Tropic Sun' sunn hemp, *Crotalaria juncea* L. Research Extension Series 036. College of Tropical Agriculture and Human Resources, University of Hawaii.

Sarkar, S.K. (2010). *Sunnhemp phyllody-* an emerging threat in sunn hemp seed crop. In: national symposium on emerging trends in pest management strategies under changing climate scenario. 20-21 December, Orissa. University of Agriculture and Technology, Orissa, India.

Sarkar, S.K. and Gawande, S.P. (2016). Diseases of Jute and allied fibre crops and their management. *J. Mycopathol. Res.*, 54(3): 321-337.

Sarkar, S.K., Hazra, S.K., Sen, H.S., Karmakar, P.G and Tripathi, M.K. 2015. Sunnhemp in India. ICAR—Central Research Institute for Jute and Allied Fibres, Barrackpore, Kolkata. Pp. 136.

Shil, S., Mitra, J. and Pandey, S.K. (2018). JRJ 610 (Prankur): A new sunn hemp (*Crotalaria juncea* L.) variety for high-yielding and superior quality fibre. *Journal of Pharmacognosy and Phytochemistry*, **7**(4): 2354-2357.

Solomon, J.J. (1972). Studies on southern sunn hemp mosaic virus. Doctoral Thesis submitted to the University of Madras, India.

Solomon, J.J. and Sulochana, C.B. (1979). Translocation of southern sunn hemp mosaic virus in *Crotalaria juncea* L. *Proc. Indian Acad. Sci. (Plant Sci.)*, 89(1): 57-60.

Wang, C.L. and Dai, Y.L. (2018). First report of sunn hemp Fusarium wilt caused by *Fusarium udum* f. sp. *crotolariae* in Taiwan. *Plant Disease*, 102(5): 1031.

Wang, K.H., Sipes, B.S. and Schmitt, D.P. (2002). "*Crotalaria* as a cover crop for nematode management: A review." *Nematropica*, 32: 35–57.

Wang, K.K. and Mcsorley, R. (2009). Management of nematodes and soil fertility with sunn hemp cover crop. *UF/IFAS Extension*, Gainsville, Florida.

White, G.A. and Haun, J.R. (1965). Growing *Crotalaria juncea*, a multi-purpose fiber legume, for paper pulp. *Econ. Bot.*, 19: 175-183.

Win, N.K., Jung, H. and Ohga, S. (2011). Characterization of sunn hemp witches's broom phytoplasma in Myanmmar. *Journal of the Faculty of Agriculture Kyushu University*.

Chapter - 12

Diseases of Rubber (*Hevea brasiliensis*) and their Integrated Management

Anwesha Sharma[1], Alinaj Yasin, Madhusmita Mahanta and Pranab Dutta[2]

School of Crop Protection
College of Post Graduate Studies in Agricultural Sciences (CPGSAS),
Central Agricultural University (Imphal), Umiam, Meghalaya-793103

Rubber (*Hevea brasiliensis*) of the family Euphorbiaceae, is commercially planted as the primary source of rubber. Other plants like *Palaquium gutta, Ficus elastica,* etc. also serve as a source of natural rubber. It is an elastic hydrocarbon polymer exuding a milky emulsion in the sap of some groups of plants. The total world coverage of para rubber *Hevea brasiliensis* is over 6 million hectares with an annual world production of 3.5 million tons. India is the 3rd largest producer of natural rubber and this rubber industry accounts for more than 12000 crores of turnover. Kerala and Tamil Nadu are the leading producers in India. The different rubber trees are affected by a number of pests and diseases causing disturbances in production (Mazlan *et al.,* 2019). The pests and diseases can cause damage and defoliation of the leaves, affects bark renewal, kills the branches and trees and also affects the roots, which may cause a huge loss in natural rubber production affecting the economy (Iyanage and Jacob, 1992; Wastie 1975).

The common pathogens attacking rubber are *Helminthosporium, Colletotrichum, Oidium, Fusarium,* etc. on the other hand, the important pests attacking rubber plants are termites, mealy bugs, scales, mites, etc. The losses caused by these pests and diseases are inevitable and need to be managed as this industry provides a major source of income for several millions of South and South-East Asian families, alone.

i. Foliar Diseases

1. Leaf Spot/ Bird's eye spot

Symptoms

The disease is characterized by a number of circular spots, small in size, scattered across the leaf surface. The spots consist of a distinct brown margin and a transparent center. The younger leaves turn black and wrinkled once affected by the pathogen, whereas the mature, older leaves bear necrotic tissues with shot-hole symptoms. In severe infection, leaves shrivel, turn black and defoliate.

Causal organism: *Helminthosporium heveae*

Epidemiology

Heavy rainfall favours the growth of the pathogen. A temperature range of 25-30°C is extremely conducive for pathogens along with a high humidity level of more than 90%. A high dose of N- based fertilizers is also favourable for pathogens to grow.

Disease cycle

Infected plant parts and debris serve as the primary source of inoculum for the disease. The secondary spread is caused by collateral hosts.

Integrated management

Cultural

- » The diseased and dead plant parts should be removed.

- » Reduced humidity level and presence of sunlight reduce leaf infection.

- » Optimum temperature and plant spacing should be maintained.

Biological

- » Application of *Trichoderma viride, T. harzianum* and *Pseudomonas fluorescens* as seed and soil application is an effective way to control *Phytophthora*.

- » Shade dried neem leaves @ 2t/ha and covering this with mud is also useful.

Chemical

» Spraying of Dithane M-45 @ (6tbsp/16 liters of water) on fully expanded leaves is highly effective.

» A weekly spray of 0.2% Zineb is also very effective.

2. Anthracnose

Symptoms

The initial symptoms show yellowish-green unhealthy leaves. The pathogen mainly attacks tender leaves of young plants starting infection from the tip towards the base. Spots appear on the leaves, circular up to 5mm in diameter. The spots are brownish with a distinct brown margin and yellow halo. The spots coalesce to form lesions at the edge which gradually move towards the middle. In severe conditions, the leaves are distorted, shrivel and wither away (Liu *et al.*, 2018)

Causal organism: *Colletotrichum gleosporioides*

Epidemiology

Wet weather condition with rainfall is extremely favourable for the growth of pathogen causing extensive leaf fall. The optimum temperature for growth and sporulation ranges from 26-32°C (Wastie, 1972), and the maximum at 28°C (Wimalajeewa, 1967).

Disease cycle

The fungus remains in the infected plant debris in the field. Soil-borne conidia serve as primary inoculum, spread by rainwater splash or splash irrigation. The secondary spread in the field is caused by air-borne conidia.

Integrated management

Cultural

» The diseased and dead plant parts should be removed.

» Optimum temperature and plant spacing should be maintained.

» Management practices should include cultural operations such as proper drainage in the area with proper nutrient treatment.

Biological

» Application of *Trichoderma viride, T. harzianum* and *Pseudomonas fluorescens* as seed and soil application is an effective way to control the pathogen.

» Shade dried neem leaves @ 2t/ha and covering this with mud is also useful.

Chemical

» Chemical practices should contain the application of Cu-based fungicide @ 2g a.i. /liter of water, 4 times at weekly intervals.

» A spray of 1% Bordeaux mixture is also very effective.

3. Powdery Mildew

The disease is widespread across many countries like India, China, Malaysia, etc. (Ramakrishnan and Pillai, 1962), but in a few countries, it is of minor importance (Peries, 1966).

Symptoms

Common symptoms of the disease include the appearance of white dusty colonies on both sides of the leaf with translucent yellow blotching. On the lower side of the leaf, circular colonies of fungal hyphae can be seen near the veins. In the younger leaves, infection leads to premature defoliation. In the semi-mature leaves, distortion is observed developing certain necrotic spots as they grow. Under severe conditions, extreme defoliation occurs resulting in loss of yield and poor bark renewal (Li *et al.*, 2016).

Causal organism: *Oidium heveae*

Epidemiology

High temperature and high humidity extensively favour the growth and viability of the pathogen. The optimum temperature favouring germination of the pathogen is 23-25°C and humidity should be more than 90%. Wet weather conditions are also favourable for the incidence of the disease.

Disease cycle

The fungus survives in the crop debris which serves as a primary source of inoculum. Secondary spread occurs through airborne conidia.

Integrated management

Cultural

» Removal of the affected leaves burning and them is a good cultural practice.

» Reducing humidity within the area of cultivation, good air circulation through the canopy, and good light exposure to all leaves and clusters, aid in managing powdery mildew.

Biological

» Applying *Trichoderma hamatum, T. harzianum* and *T. viride* are the most promising methods.

» Spraying neem oil, and canola oil, reduces powdery mildew severity in the plants.

Chemical

» Application of Sulfur dust @ 5-7 days intervals is highly efficient in controlling the disease, especially at the young stage.

» Spraying of Bavistin @1g/litre of water is also effective.

4. Corynespora leaf fall

This disease is highly prevalent in Southeast Asia and African countries.

Symptoms

The disease is characterized by greyish brown spots with railway track symptoms. The typical symptoms of the disease are brown elongated spots on both younger and older leaves. The spots are surrounded by a yellow halo. The greyish-black lesions can also be seen on the petioles. The infection causes premature leaf fall and shoots die back. Infection leads to prolonged immaturity and the death of plants.

Causal organism: *Corynespora cassiicola*

Epidemiology

The high temperature from 25-30 °C is extremely favourable for the pathogen. High humidity with leaf wetness @ 90% is suitable for the pathogen (Manju, 2011).

Disease cycle

Primary spread occurs through plant debris. Secondary infection is through wind-borne conidia. Several alternate hosts of plants serve as a source of pathogen overwintering.

Integrated management

Cultural

> » Removal of the affected leaves and burning them is a good cultural practice.

> » Artificial defoliation to avoid leaf fall.

> » Optimum plant spacing should be maintained.

Biological

> » Applying *Trichoderma hamatum*, *Trichoderma harzianum* and *Trichoderma viride* are the most promising methods.

> » Spraying neem oil, and canola oil, as biological control, in the plants.

Chemical

> » Spraying Mancozeb @ 2.5% as foliar spray is effective.

> » Spraying @4.5 kg Cu/ha of copper oxychloride or Copper oxide before monsoon.

> » In India 2 sprayings with 1% Bordeaux mixture or 0.2% Zineb are recommended.

5. Leaf blight or Leaf fall

The disease is predominantly observed in India, Brazil, and other South East Asian countries like Thailand, Malaysia, Vietnam, etc.

Symptoms

The mature leaves are covered with brown and yellowish, circular, water-soaked lesions. The lesions often coalesce together forming larger necrotic areas. The budded seedlings show vascular discolouration. Along the petiole, some greyish black lesions develop along with white globules of coagulated latex, at the point of entry of the pathogen. The formation of the abscission layer may result in premature defoliation (Krishna *et al.*, 2018).

Causal organism: *Phytophthora palmivora* (Butl.) (Tucker, 1931), *Phytophthora heveae* Thompson (Thompson, 1929)

Epidemiology

Prolonged rainfall favours pathogen growth. The optimum temperature for the growth of pathogens ranges from 25-28°C. The pathogen survives in water and propagates. High humidity also helps in the growth of the pathogen.

Disease cycle

The fungus survives in infected debris as well as soil as chlamydospores mainly. Secondary spread is through water splash or wind-blown droplets containing spores.

Integrated management

Cultural

» The diseased and dead plant parts should be removed.

» Adequate drainage should be provided to stop water stagnation as the pathogen survives in water.

» Injury to the root system during practices like digging should be avoided.

» Reduced humidity level and presence of sunlight reduce leaf infection.

» Irrigation during the cold weather period should be maintained to avoid pathogen build-up.

Biological

» Application of *Trichoderma viride, Trichoderma harzianum* and *Pseudomonas fluorescens* as seed and soil application is an effective way to control *Phytophthora*.

» Shade dried neem leaves @ 2t/ha and covering this with mud is also useful.

Chemical

» Slightly mature seedlings from fields should be selected and soaked in

Streptocycline 500 ppm + Bordeaux mixture 1% solution for 30 minutes.

» Application of 150 kg N/ha/year through neem cake (75 kg N) and 100 kg P_2O_5 and 50 kg K_2O in 3 split doses, first dose at 15 days after growth and second dose and the third dose should be given at 40-45 days interval.

» Copper oxychloride 50% WP @ 1 Kg in 300-400 l of water/acre can also be applied.

ii. Stem and Trunk Diseases

1. Pink Diseases

Symptoms

A silky white mycelium develops on the stem like a cobweb-like film. Some pustules and necators can also be observed. The barks in the stem are covered with open wounds. The fork region of the tree or the region with trapped moisture shows salmon-pink incrustations (Akrofi *et al.*, 2014).

Causal organism: *Corticium salmonicolor* B. & Br.

Epidemiology

The optimum temperature for growth of the pathogen is 28°C and has high relative humidity. Basidiospore formation increases with rainfall. In dry conditions, it becomes unfavourable for the pathogen to remain alive.

Disease cycle

The pathogen produces pycnidia and survives on the surface of lenticels or barks. Spores of the fungus are spread by wind and rain. Infection can be made through healthy barks as well.

Integrated management

Management practices should include the removal of weeds and reduction of humidity after a long rainfall. Seedlings should be planted in full sunlight. The diseased and dead parts should be completely eradicated to reduce inoculum. Proper plant spacing should be maintained always.

The application of fungicide should be done in the infected regions after scrapping off the fungal growth wherever necessary. Organic foliar fertilizers can be applied rich in potassium to stimulate bark regeneration. Bordeaux paste can be applied on cut areas or spray 1% Bordeaux mixture.

2. Black Stripe and Stem Cracking

Symptoms

Common symptoms of the disease include some blackish stripes on pared off barks in the stems. The barks finally show cracks, which start to bleed and decay off.

Causal organism: *Phytophthora palmivora*

Epidemiology

Prolonged rainfall favours pathogen growth. The optimum temperature for the growth of pathogens ranges from 25-28°C. The pathogen survives in water and propagates. High humidity also helps in the growth of the pathogen.

Disease cycle

The fungus survives in infected debris as well as soil as chlamydospores mainly. Secondary spread is through water splash or wind-blown droplets containing spores.

Integrated management

Cultural

» The diseased and dead plant parts should be removed.

» Adequate drainage should be provided to stop water stagnation as the pathogen survives in water.

» Injury to the root system during practices like digging should be avoided.

» Reduced humidity level and presence of sunlight reduce leaf infection.

» Irrigation during the cold weather period should be maintained to avoid pathogen build-up.

Biological

» Application of *Trichoderma viride, Trichoderma harzianum and Pseudomonas fluorescens* as seed and soil application is an effective way to control *Phytophthora*.

» Shade dried neem leaves @ 2t/ha and covering this with mud is also useful.

Chemical

» Slightly mature seedlings from fields should be selected and soaked in Streptocycline 500 ppm + Bordeaux mixture 1% solution for 30 minutes.

» Application of 150 kg N/ha/year through neem cake (75 kg N) and 100 kg P_2O_5 and 50 kg K_2O in 3 split doses, first dose at 15 days after growth and second dose and the third dose should be given at 40-45 days interval.

» Copper oxychloride 50% WP @ 1 Kg in 300-400 l of water/acre can also be applied.

» Application of Dithane M-45 or Ridomil at 14 days intervals after latex collection is a successful chemical management strategy.

iii. Root Diseases

1. White Root Rot

The disease was first reported in Singapore. It is one of the most important root diseases in Asia and Africa.

Symptoms

The foliar parts show green discolouration and gradually turn yellow. Dieback can be seen in severe cases. The disease is also characterized by pre-mature flowering and fruiting due to root rotting. The tap roots are destroyed by the pathogen. Rhizomorphs undergo fructification. White rhizomorphs of fungus which are heavily branched can be seen in the exposed collars and roots infected by the fungus (ISC 2018).

Causal organism: *Rigidoporus lignosus*

Epidemiology

Wet weather is favourable for fructification. Planting trees closer also increases disease severity. High soil pH is favourable for the pathogen.

Disease cycle

The pathogen survives on dead wood or the living or dead root debris. Secondary spread is through wind-borne spores. The pathogen attacks stumps after falling.

Integrated management

Cultural

» Management practices should include complete eradication of diseased and dead-parts to reduce inoculum.

» Eradication of infected roots should be practiced.

» Proper plant spacing should be maintained always.

» Isolation trenches can be made in matured plants.

Biological

» *Trichoderma spirale, T. harzianum, Hypocrea virens, Hypocrea jecorina* etc. are extremely successful in successful control of *Rigidoporous*.

» Plantation of creepy legumes cover should be done.

Chemical

» The application of fungicide should be done in the infected areas.

» Applying Sulphur@ 150-200g/plant is extremely beneficial for all root disease control.

2. Brown Root Rot

The disease was first reported in Sri Lanka. It is one of the most important root diseases in India and Sri Lanka.

Symptoms

Chlorosis of leaves occurs and they fall off. Die back of twigs also occurs along with some dark brownish fructification. The roots consist of fungal layering, with a tan brownish colour and some zigzag lines on them. The roots turn rough with adhered soil layer. The colour of roots turns dark brown with time with honey combing (Brooks 2002).

Causal organism: *Phellinus noxius*

Epidemiology

Higher disease incidence is observed in light soil. Soil with lower moisture content is extremely favourable for the pathogen.

Disease cycle

Stumps of trees on plantations and underground root contact between healthy and diseased serves as the primary source of infection.

Integrated management

Cultural

» Eradication of infected roots should be practiced.

» Proper plant spacing should be maintained always.

» Plantation of creepy legumes cover should be done. Isolation trenches can be made in matured plants.

Biological

» Application of *Trichoderma viride, Trichoderma harzianum* and *Pseudomonas fluorescens* as seed and soil application is an effective way to control *Phytophthora*.

» Shade dried neem leaves @ 2t/ha and covering this with mud is also useful.

Chemical

» The application of fungicide should be done in the infected areas.

» Applying 2.5 tones/ha of lime is very effective.

» Applying Sulphur @ 150-200g/plant is extremely beneficial for all root disease control.

Future Prospects

With increasing disease incidence in rubber, acoounts for more economic losses which hampers the market and global trade. The rubber industry is one of the most crucial ones for developing a country's economy. As such a small loss in yield due to a pathogen attack affects the entire productivity and economy. Several modern scientific techniques could be useful for developing future elite land races, tolerant breeds and their improvement programmes. Also, efforts should be made to the characterization of the available landraces which could be useful for resolving similar problems. Biotechnological tools like chromatography, NMR, and other functional genomics techniques could be explored to find out new compounds with active potential from this unexplored plant species, development of disease-resistant high-yielding varieties, etc. Modern-day tools like nanotechnology can also be used for the long-term, effective and environment friendly control of pathogens.

Conclusion

Considering the deteriorating impact of the pathogens causing so many diseases in rubber cultivation and the resulting economic loss incurred, unravelling all possible management strategies and interactions between host and pathogen need to be given the most importance. It is to be mentioned that lots of study areas in diseases and their management strategies still remain unexplored. The plant host defense system needs to be understood and developed for tolerance along with integrated management strategies.

References

Akrofi, A. Y., Amoako-Atta, I., Assuah, M. and Kumi-Asare, E. (2014) Pink Disease Caused by *Erythricium salmonicolor* (Berk. & Broome) Burdsall: An Epidemiological Assessment of its Potential Effect on Cocoa Production in Ghana. *J. Plant Pathol. Microb.* 5 (215). doi:10.4172/2157- 7471.1000215.

Brooks, F. E. (2002). Brown root rot disease in American Samoa's tropical rain forests. *Pac. Sci.* 56: 377-387.

Iyanage, A. S. and Jacob, C. K. (1992). Diseases of economic importance in rubber. In: *Developments in Crop Science*, (Eds. Sethuraj, M. R. Mathew, N.M.), Elsevier, Netherlands, pp. 325-359.

Invasive Species Compendium. (2018). *Rigidoporus microporus* (white root disease of rubber). Retrieved 2021, from CABI Invasive Species Compendium: https://www.cabi.org/isc/datasheet/47610.

Krishnan, A., Joseph, L. and Roy, C. B. (2018). An insight into *Hevea-Phytophthora* interaction: The story of *Hevea* defence and *Phytophthora* counter defence mediated through molecular signalling. *Curr. Plant Biol.* doi: 10.1016/j.cpb.2018.11.009.

Liu, X., Li, B., Cai, J. *et al. Colletotrichum* Species Causing Anthracnose of Rubber Trees in China. *Sci Rep* **8**: 10435. doi: 10.1038/s41598-018-28166-7

Li, X., Bi, Z., Di, R., Liang, P., He, Q., Liu, W. and Zheng, F. (2016). Identification of powdery mildew responsiveness genes in *Hevea brasiliensis* through mRNA differential display. *Int. J. Mol. Sci.* 17(2), 181.

Mazlan, S., Md, N., Wahab, A., Sulaiman, Z, Rajandas, H. and Zulperi, D. (2019). Major Diseases of Rubber (*Hevea brasiliensis*) in Malaysia. *Pertanika J. Sch. Res. Rev.* 5(2): 10-21.

Manju, M. J. (2011). Epidemiology and Management of Corynespora leaf Fall Disease of Rubber caused by *Corynespora cassiicola* (Berk & Curt.) Wei. *Ph.D. Thesis*, University of Agricultural Sciences, Dharward, India.

Peries, O.S. (1966). Present status and methods of control of leaf and panel diseases of Hevea in South East Asian and African countries. *J. Rubber Res. Inst.*, 42: 35-47.

Ramakrishnan, T.S. and Radhakrishna Pillay, P.N. (1962). Powdery mildew of rubber. *Rubber Board Bull.*, 5: 187.

Thompson, A. (1929). Phytophthora species in Malaya. *Malay. Agric. J.*, 17: 53-100.

Tucker, CM. (1931). Taxonomy of the genus Phytophthora de Bary. Research Bulletin of the University of Missouri Agricultural Experimental Station, 1953, pp 208.

Wastie R. L (1975) Diseases of Rubber and their Control. *PANS Pest Articles News Summ.* 21(3): 268-288.

Chapter - 13

Diseases of Cardamom (*Elettaria cardamomum*) and their Integrated Management

Diganggana Talukdar

College of Horticulture, Bermiok,
Sikkim (Central Agricultural University, Imphal)

Introduction

Cardamom (*Elettaria cardamomum*) also known as "Queen of Spices" belonging to the family Zingiberaceae and order Scitaminae is one of the most important spice crops cultivated in the different parts of India and overseas. The cardamom is of two types, small and large cardamom (Talukdar and Anand, 2021; Pathak, 2021) both have a very high market value all over the world. In India, small cardamom is mainly cultivated in the southern states of India mainly Kerala (60%), Karnataka (30%), and Tamil Nadu (10%). On other hand, large cardamom is one of the most important spice crops cultivated in the eastern part of India, i.e., sub-Himalayan states of Sikkim, Darjeeling, Nagaland and Arunachal Pradesh (Pathak, 2021) and its been reported that the production of large cardamom ranged from 4500 to 5000 metric tons annually mostly covering the area of Sikkim state and the Darjeeling district of West Bengal. Morphologically, cardamom is a perennial herb consisting of subterranean rhizomes and several leafy aerial tillers. The number of such rhizomatous leafy shoots varies between 15 and 140 inches in a single plant. The fruit is a round or oval-shaped capsule with several seeds. The capsule wall is reddish-brown to dark pink in colour and has little rough textures. Seeds are white when immature and become bigger and bolder than the normal cardamom seeds. Large cardamom is bestowed with the pleasant aromatic odour for which it is extensively used as a flavouring agent for food preparations in India. Besides being used as spices for its aroma, large cardamom also has several medicinal values like disinfecting teeth and gums infections, considered as an antidote to either snake venom or scorpion venom, preventive as well as curative measures for throat troubles, and congestion of the lungs, inflammation of eyelids, digestive disorders and in the treatment of pulmonary tuberculosis (Pathak, 2014). Among most of the factors, diseases proved to be the prime factor for the reduction of the productivity of this crop (Paudel, et al., 2018). These diseases have spread due to drastic changes in the ecosystem, inadequate rain in dry months, and the absence of good agricultural practices by the farmers. Many cardamom farmers failed to plant varieties suitable to their altitude (Pathak, 2014). The fifteen most highly infected diseases of Cardamom have been depicted in this chapter.

1. Capsul rot or Azhukal disease

Etiology: *Phytophthora* spp. belonging to the oomycetes group of fungi being soil borne in nature.

Disease symptoms

The disease appears during the rainy season. On the infected leaves, water-soaked lesions appear first followed by rotting and shedding of leaves along the veins. The infected capsules become dull greenish-brown and decay. This emits a foul smell and is subsequently shed. The infection spreads to the panicle and tillers resulting in their decay (Nair, 2020).

Epidemiology

The disease spreads through soil, water and wind. Moreover, continuous rainfall and high relative humidity exaggerate the disease infection.

Management

Following the common, cultural and mechanical practices help in reducing diseases. In the biological control, *Trichoderma harzianum* 0.50% WS @ 100 g/plant (as soil treatment) is very effective. Application of 100 g product/ plant along with neem cake (0.5 Kg/plant) and 5 Kg FYM/plant also helps in reducing the disease. Soil drenching with fosetyl-AL 80% WP @ 900-1200 g in 300-400 l of water/acre helps in plummeting the disease (Belbase *et al.*, 2018).

2. Primary nursery leaf spot

Etiology: This disease is caused by the fungus *Phyllosticta elettariae*

Disease symptoms

The disease appears as small round or oval spots, which are dull, and white in colour. With time, these spots turn necrotic and develop a shot hole at the center. Sometimes, the spots may be surrounded by water-soaked areas on the leaves. Disease intensity upsurges in open nurseries that are exposed to direct sunlight. In such conditions, abundant spots develop on the leaf areas. (Satyagopal *et al.*, 2014).

Epidemiology

Under direct sunlight, if the nurseries get exposed, the disease intensity gets augmented in the open nurseries. High humidity or persistent dew rains are favourable for this disease. It appears mostly during the months of February to April with the reception of summer showers (Satyagopal *et al.*, 2014).

Management

Under cultural control, raising a nursery in fertile soil is the prerequisite. Avoid direct sunlight on nursery beds. but Using a light shade net is recommended. Early sowing of seeds in August-September will ensure mature seedlings which are less prone to diseases during the southwest monsoon. Under biological control, the application of *Trichoderma viride* or *Trichoderma harzianum* and *Pseudomonas fluorescens* as seed/seedling/planting material, nursery treatment and soil application is very fruitful. In chemical control spraying or drenching the soil after germination of seedlings with copper oxy-chloride @ 1 g in 300-400 l of water/acre is recommended (Annonymous, 2014a).

3. Rhizome rot

Etiology: This disease is caused by several soil-borne fungi namely *Rhizoctonia solani*, *Pythium* spp. and *Fusarium* spp. (Karkee and Mandal, 2020; Balbase *et al.*, 2018).

Symptoms

Chlorosis of older leaves, splitting at the base of the pseudostem, more mature plants may exhibit a brown to black discolouration and decay of the rhizome and decay causes the death of plants and the roots get blackened. Brown discolouration in the colour region indicates infection by *Rhizoctonia*, pinkish discolouration and softness of pseudostem indicates Pythium, and sudden wilting of plants indicates infection by Fusarium (Karkee and Mandal, 2020; Talukdar and Anand, 2021).

Epidemiology

This complex disease is all soil-born in nature. Infected rhizomes also play a major in the spreading of the pathogens. Rhizome rot of cardamom is directly related to soil moisture, atmospheric humidity and number of propagules in the soil, rainfall, and number of rainy days during the period coupled with low soil and ambient temperature (Karkee and Mandal, 2020; Balbase, *et al.*, 2018).

Management

Always plant the cardamom plants in well-drained soils. Drenching the soils with bio-fungicides (eg: *Trichoderma*) prior to planting in the nurseries is very fruitful. Remove the infected clumps including rhizomes whenever observed. Treatment of seeds with *Trichoderma* culture (50 ml spore suspension for 100 g of seed) is desirable as a prophylactic measure for managing nursery rot diseases (Anonymous, 2019; Talukdar and Anand, 2021). Collecting and burying the affected flowers or spikes as a mechanical control measure is very effective (Anonymous, 2014b). But the plants affected by the viral diseases cannot be cured but the losses can be minimized by adopting appropriate management practices like keeping a constant vigilance for detecting disease-affected parts and uprooting and destroying affected plants as soon as symptoms appear. Seedlings must be procured from certified nurseries and propagation through suckers must be done only through certified multiplication nurseries (Pathak, 2021). The suckers deep in *Trichoderma* or *Pseudomonas fluorescence* or *Bacillus subtilis* @ 5g/ L water for 1 hour and then transplanting is very effective (Balabse, *et al.*, 2018). Karkee and Mandal, (2020) suggested some cultural approaches to monitoring this disease and they are like the nursery raised in fertile soil, early sowing of suckers from August to September, collecting and destroying disease-infected plant parts, providing proper irrigation at critical stages of the crop, avoiding waterlogging conditions, regulating shades in densely shaded areas, avoid water stress during the flowering stage and destroying the alternate or collateral hosts such as ginger, turmeric and castor close by.

4. Cercospora leaf spot or Secondary nursery leaf spot

Etiology: This is a fungal disease caused by Cercospora *zingiberi*

Disease symptoms

This disease is prevalent in the nursery areas and plantations orchards which are diagnosed as rectangular muddy red stripes running along the veins of the leaves. This disease is mainly scattered as spherical blotches on the leaves. In the beginning, the small lesions or spots measure a few mm and with time, these several spots coalesce to cover larger areas. Leaf rust is often perceived on the older leaves as whitish powdery pustules on the undersurface of the leaves with yellowish necrotic patches. While on the upper surface, diseased leaves rusty appearance is observed (Annonymous, 2014a).

Epidemiology: Disease appears mostly during February- April months with the receipt of summer showers.

Management

Under cultural control, cardamom seeds should be sown in the month of August to September in order to ensure sufficient seedling growth so that they develop the capacity to tolerate the disease incidence. Under chemical control, spraying or drenching of the soil with Copper oxychloride 50%WP @ 1kg in 300 to 400 l of water per acre is found to be effective (Annonymous, 2014a).

5. Leaf blotch

Etiology: Leaf blotch disease of small cardamom is caused by the fungal pathogen *Phaeodactylium alpiniae* and it is generally considered a minor disease occurring during monsoon months (Thomas and Bhai, 2002).

Symptoms

The leaves of the diseased plants form large spots of irregular lesions along with alternate shades of light and dark brown necrotic areas which are mainly detected on mature leaves. Later on, grey-brown fungal mycelium and spores develop as masses on the underside of these lesions (Ajay *et al.*, 2013).

Epidemiology

This disease appears with the onset of the monsoon season. High rainfall and humidity are very conducive to the emergence of this disease (Satyagopal, *et al.*, 2014).

Management

Under cultural control, avoid waterlogged conditions. The application of Bordeaux mixture (1.0%) followed by Propiconazole (0.1%) is effective in combating the disease incidence. The combination of fungicides, Trifloxystrobin + Tebuconazole

(1.0%), Thiophanate Methyl (0.2%), Mancozeb (0.2%) and Tebuconazole 0.1%) were also effective in controlling the disease (Ajay *et al.*, 2013; Satyagopal, *et al.*, 2014).

6. Leaf blight or Chenthal disease

Etiology: This disease is caused by the fungus *Colletotrichum gloeosporioides*

Symptoms

The lesions on the pseudostem become necrotic, and as a result, entire leaves dry out, giving the plant a burned appearance. The infection first innates on the young middle-aged leaves in the form of elongate to ovoid, brown-coloured patches which rapidly convert into necrotic and dry lesions mostly observed on leaf margins. In severe cases, the entire leaf area on one side of the midrib is affected. Leaf blight or drying of leaves in patches is observed during October to February months (Annonymous, 2014). Saju *et al.*, 2010 and Satyagopal, *et al.*, (2014) also reported that the characteristic symptoms include light brown irregular spots at the initial stage starting along the margins of the leaves which later on extend to the midrib and turn into reddish brown lesions and with the length of time, the leaves get dried.

Epidemiology

Environmental and human factors are more adverse in the large cardamom-growing pockets of Sikkim. As per farmers, representation, the planting materials of the cultivar, would have carried the inoculum, or else the alien/new cultivar would have been more susceptible to the native pathogen. Highly favourable climatic conditions developed later would have helped the pathogen inoculum to increase and invade other varieties in various locations too. Hailstorm injury results in the drying of leaves and from time to time weak foliar pathogens enter through the wounds and cause the blighting of the plant (Saju, *et al.*, 2010).

Management

» The incorporation of biocontrol agent *Trichoderma* multiplied in the suitable organic medium in the plant base (1 kg per clump) prior to the onset of monsoon season controls this clump rot disease effectively.

» Usage of Bordeaux mixture 1 %, when found necessary may be applied.

» Regular rouging of virus-affected plants must be accomplished to reduce the spread (Satyagopal, *et al.*, 2014).

» Later on, the rouged plants should be destroyed by burning (Anonymous, 2019; Saju *et al.*, 2010; Balbase, *et al.*, 2018) to avoid traces of the virus.

7. Leaf rust

Etiology

This disease is caused by the fungus *Phakospora elettariae* and being heterocyclic in nature, it produces both uredospores and teliospores. The spores are uredospores that are sessile, obovoid to ellipsoidal, echinulate, yellowish brown and become cinnamon brown with age. Teliospores are one-celled, irregularly arranged in 2-7 layers, oblong to ellipsoidal (Satyagopal, *et al.*, 2014).

Symptoms

The disease is observed as minute, brown uredospores on the lower surface of leaves. The uredospores are surrounded by chlorotic halos during the early stages of development. In severe cases, the whole leaf gets covered by uredospores resulting in premature drying (Satyagopal, *et al.*, 2014. Leaf rust is often seen on the mature leaves as whitish powdery pustules on the undersurface of the leaves with corresponding yellow necrotic patches on the upper surfaces. Disease leaves show a rusty appearance on it the same as the disease depicts (Anonymous, 2021).

Epidemiology

The disease is frequently detected during the months of May to June on the lower surface of leaves.

Management

Organic management is effective in rust disease. Spraying the crop with Mancozeb (0.25%) or Tebuconazole (0.05%) and repeating after 10 to 15 days intervals proves to be effective in controlling the disease

8. Root tip rot/leaf yellowing/pseudostem rot/stem lodging/ panicle wilt

Etiology: It is caused by soil-borne fungi *Fusarium oxysporum*

Symptoms

Root tip rot and leaf yellowing occur widespread in several cardamom plantations in the Idukki district of Kerala (Satyagopal, *et al.*, 2014). The symptoms are yellowing of the foliage resulting in leaf drying. Usually, the symptoms start from the older basal leaves onwards and reach towards the middle portion of the tillers. The earlier affected basal leaves become fully yellow and soon dry off. Characteristic visible symptoms are seen in the root system also. In the affected plants, root tip portions show indications of decay that proceeds gradually towards the plant base. Such roots show shrivelling and an off-white to grey colour at the root tips. If sufficient moisture is present in the soil, the infected plant parts show decay.

Epidemiology

The disease makes its appearance after the monsoon rains and becomes severe during the summer months (Satyagopal, *et al.*, 2014).

Management

Integrated disease management strategies using plant sanitation and fungicides and eco-friendly systems involving the use of bioagents and beneficial microorganisms such as *Trichoderma* and *Pseudomonas* are being recommended. Researchers stated that the usage of bioagents such as *Trichoderma, Bacillus* sp, *Pseudomonas, etc* for the disease management in cardamom has been proved to be effective. Vijayan and Thomas, (2009) isolated the two indigenous strains of *Trichoderma* viz. ICRI isolated T12 (*Trichoderma harzianum*) and T14 (*T.viride*) from the cardamom soils which were found to be very effective in the management of cardamom soil-borne diseases. VAM inoculation at the seedling stage significantly enhanced the growth and vigour of the seedlings. According to Satyagopal, *et al.*, (2014), under field control trials using systemic chemical fungicides viz., Carbendazin (0.2 per cent), Hexaconazole (0.2 per cent) or Thiophanate methyl (0.2 per cent), it showed that *Fusarium oxysporum* disease infections in small cardamom can be brought under control when it is applied as spraying or soil drenching. Phytosanitation by pruning dry leaves, uprooting, and removal of diseased tillers, panicles, and rhizomes may be carried out. Destroy them by burning or deep burying them. The top priority must be inculcated under the phytosanitary operations to reduce the pathogen inoculum load in the cardamom plantations to maintain the ecological balance in the environment. Subsequently, soil drenching and spraying should be carried out thrice at monthly intervals using chemical fungicides such as carbendazim (0.2 per cent), thiophanate methyl (0.2 per cent), or hexaconazole (0.2 per cent) in the plantations for management of Fusarium diseases. Three rounds of applications may be given during the August, September, and October months. Root rot-affected plantations may be given foliar sprays with 1-2 per cent DAP depending on the age of the plant. Apply biocontrol agents such as 1 per cent *Trichoderma harzianum* (10^9 CFU/ml) as basal application and 1 per cent *Pseudomonas fluorescens* (10^9 CFU/ ml) as spray and soil drenching after 10 days of fungicidal application. The application of biocontrol agents must be applied twice in the cardamom plantations.

9. Leaf streak

Etiology: Leaf streak is caused by *Pestalotiopsis royenae*. Accervuli is dark brown to black. Conidia are usually multiseptate, 5-celled where the three median ones are brown in colour and two end ones are hyaline, having simple or branched appendages (Vijayan *et al.*, 2019).

Symptoms

The diagnostic symptoms of leaf streaks are characterized by the presence of abundant elongated translucent streaks on young leaves along the veins as the name depicts. After 3 to 4 days these streaks turn reddish brown and develop a straw-coloured necrotic area at the center which is surrounded by dark brown margins.

Epidemiology

The cool and moist weather favours the disease. The pathogen overwinters in the form of acervuli or dormant mycelium during off-seasons.

Management

Fungicides like Carbendazim (0.1%), and Zineb (0.25%) are effective in controlling the disease.

10. Phoma leaf spot disease

Etioligy: Leaf spot caused by *Phoma hedericola.*, a fungus (Saju *et al.*, 2011), was found to be of serious concern in the seedling nurseries in Arunachal Pradesh and field plants in Sikkim.

Symptoms

Phoma disease is characterised by numerous water-soaked lesions which are round in shape that appear on the lamina. Later on, these lesions coalesce and the leaves become yellowish and dry out completely (Vijayan *et al.*, 2019).

Epidemiology

This disease spreads rapidly during continuous rain and consequent damage indicates its potential to devastate the whole plantation. In Sikkim, the disease was found to occur during late winter and peak rainy periods.

Management

Cultural control with field phytosanitation by removal and destruction of disease-affected plants or plant parts is very effective in controlling this disease. The provision of adequate drainage is a must. Spraying of 1% Bordeaux mixture at 20-25 days intervals during rainy days based on disease severity is recommended under chemical control (Vijayan *et al.*, 2019).

11. Katte disease

Etiology

Katte disease is caused by the Cardamom mosaic virus (CdMV) which restricts the production in all cardamom-growing regions of the world. In the present study by Tiwari *et al.*, (2016), among 84 cardamom plantations in 44 locations of Karnataka and Kerala, the disease incidence ranged from 50% to 85%.

Symptoms

According to Vijayanandraj *et al.*, (2017), the first visible symptom appears as spindle-shaped slender chlorotic flecks measuring 2-5 mm in length on the youngest leaf of the affected tillers. Later on, these flecks develop into pale green discontinuous stripes on the leaves. The stripes run parallel to the vein from the midrib to the leaf margin. All the subsequently emerging new leaves show characteristic mosaic symptoms with chlorotic and green stripes. As the leaf matures, the mosaic symptoms are more or less masked. The disease is systemic in nature and it gradually spreads to all the tillers in a clump. Younger plants express symptoms earlier than grown-up clumps. Katte-infected plants can survive for many years which serves as sources of inoculum. If the plants are infected by this virus at the seedling stage the will be a total loss of the total plantations. In bearing clumps, the loss ranged up to 68% within three years after infection.

Epidemiology

The primary source of inoculum is the plant infected itself as it can survive for several years even after the katte infection. The aphid vector *Pentalonia caladii* (formerly *P. nigronervosa* f.sp. *caladii*) transmits this virus. It is also transmitted by infected rhizomes, infected clones, seedlings raised near infected plantations, volunteer plants, and a few of the infected Zingiberacae plants (*Venugopal, 2002*). The aphids play an important role in the disease spread as they are prevalent throughout the year in plantations, except during monsoon season (Vijayanandraj, *et al.*, 2017).

Management

Chemical control: Chemical control measures are considered to be less effective in managing viral diseases owing to the non-persistent and semi-persistent modes of transmission of viruses by the vectors. Spraying recommended insecticides after undertaking trashing operation increases the efficacy of application and manages the vector to a greater extent. In an insecticidal trial, katte disease incidence was found to be significantly higher in the treatments where phorate, dimethoate, phosphamidon, carbofuran and quinalphos were applied, than in the unsprayed treatments indicating that insecticidal application may not be effective in preventing the spread of the disease if the source of inoculum in maintained in the plantation.

Removal of breeding sites

Timely removal of disorientated old parts and collateral hosts like Colocasia and Caladium effectively reduces the aphid population and checks the succeeding spread of viral diseases.

Biopesticides

Botanicals like Neem extracts at 0. 1 % concentration was found to reduce the population of aphids on cardamom and was lethal at higher concentrations. Aqueous extracts of *Acorus calamus* (dried rhizome), *Annona squamosa* (seeds) and *Lawsonia inermis* (leaves) were also found to reduce the setting percentage of aphids on cardamom (Baju and Bhat, 2012).

12. Chirkey disease

Etiology

Chirkey disease is a viral disease caused by the Mosaic Streak Virus consisting of polyhedral particles measuring 40 nm diameters. (Paudel, *et al.*, 2018).

Epidemiology

The corn aphid *Ropholosiphum maydis* is the vector for the disease. The disease is also spread by planting infected suckers.

Symptoms

This is one of the major diseases of large cardamom which spreads swiftly and the symptoms are distinguishable unless it is carefully observed. The disease is characterized by the mosaic appearance on leaves. The symptom is more protuberant on newly emerged leaves where discrete pale green to yellow longitudinal strips can be seen that is running parallel to each other. Here characteristic mosaic symptoms are also observed due to the formation of 2.5mm flecks on the leaves. The flowering is significantly reduced thus reducing the yield of the crop (Bhat, *et al.*, 2020; Talukdar, 2021).

Management

In case of Removal of weeds. Use healthy planting material. Control aphid vectors. Treatment of planting materials with *Pseudomonas fluorescens* @0.5%. (Mandal, *et al.*, 2012; Talukdar, 2021). Commercially available formulation of entomopathogenic fungi (*Beauveria bassiana*) Mycotrol or Biosoft @ 3.0g/l, Vertalec or Inovert or Biocatch (*Verticillum lecanii*) @ 3.0g/l, Prioroty (*Paecelomyces fumosoroseus*) @2.5ml/l should be applied (Paudel, *et al.*, 2018). Time to time inspection of the diseased plants with the utilization of certified virus-free plants is essential for the management of chirkey disease. Early identification of the diseased plants, uprooting and burning help reduce spread the spread of the viral diseases (Talukdar and Anand, 2021). Avoid collecting planting materials from infected gardens. Establish nurseries about 500 m away from the plantation in order to avoid aphid colonization (Anonymous, 2019).

13. Foorkey disease

Etiology

It is caused by an isometric virus particle of 17 to 20 nm named Cardamom bushy dwarf virus (Annonymous, 2014 b, Mandal, *et al.*, 2007; Paudel, *et al.*, 2018; Vijayanandraj *et al.*, 2017)

Epidemiology

Transmitted through banana aphid *Pentalonia nigronervosa*. Diseased plants survive for a few years but remain sterile and unproductive.

Symptoms

The symptoms of the disease are the production of bushy growth of stunted shoots which are sterile. The affected plants produce profuse stunted shoots which fail to produce flowers (Paudel, *et al.*, 2018). The leaves become small, lightly curled and pale green in colour. The inflorescence becomes stunted, thereby producing no flowers and fruits. The capsule size was reduced and chaffy without seeds (Talukdar, 2021).

Management

» Application of commercially available formulation of insect pathogenic fungi (*Beauveria bassiana*) Mycotrol or Biosoft @ 3.0g/l, Vertalec or Inovert or Biocatch (*Verticillum lecanii*) @ 3.0g/l, Prioroty (*Paecelomyces fumosoroseus*) @2.5ml/l proved to be very effective (Mandal, *et al.*, 2007; Mandal, *et al.*, 2012; Paudel, *et al.*, 2018) to control the aphids which are the real threat for the spread of the disease.

» Uprooting infected plants must be accomplished as a cultural method (Talukdar, 2021).

» The surrounding areas must be drenched with systemic bioinsecticides to control the movement of aphids on the main host as well as collateral hosts such as banana, peach and squash (Talukdar, 2021).

» The diseased plants including rhizomes must be uprooted and destroyed as and when they are traced. Then, the uprooted plants must be taken to an isolated place and chopped into small pieces buried in deep pits for their quick decomposition. Usage of healthy and disease-free planting host plants of the aphids in and near the plantation is a must (Annonymous, 2014b; Talukdar and Anand, 2021).

14. Cardamom necrosis/Nilgiri necrosis; Cardamom necrosis virus

Etiology

It is caused by Cardamom necrosis virus which is flexuous rod-shaped (570-700 nm length and 10-12 nm) breadth morphologically. This disease was first noticed in a severe form in Nilgris, Tamil Nadu and hence the name Nilgiri necrosis disease was derived (Venugopal, 1995). Gradually it spread to the new pockets in Kerala and Tamilnadu and a few areas in Karnataka.

Disease symptoms

Young leaves exhibit whitish to yellowish continuous or broken streaks proceeding from the midrib to the leaf margins and later turn reddish brown. Often leaf shredding is noticed. The affected plants are stunted and fail to bear the panicles and capsules. Irregular yellowish and necrotic patches lesions are seen on younger leaves. A stunted appearance is exhibited on the Nilgiri necrosis-affected plant (Anonymous 2014a; Anonymous 2019)

Epidemiology

The disease is not transmitted by seed, soil, sap and mechanical means and no insect transmission of the disease from infected to healthy plant was recorded. The disease is transmitted through infected rhizomes (Venugopal, 1995).

Management

According to Anonymous, (2019) several integrated management practices proved fruitful for effective management. Under common cultural practices, Collecting and destroying crop debris is a must. Provide irrigation at critical growth stages of the crop but avoid waterlogging. Regulate shade in thickly shaded areas and avoid water stress during the flowering stage. Enhancing parasitic activity by the naturally available parasites and predators to reduce chemical spray, when 1-2 larval parasitoids are observed. Destroy the collateral and alternate hosts such as castor and spices like ginger and turmeric in the nearby areas. Maintain optimum plant density always. Always fill the gaps with healthy disease-free materials. Mulching the plant basins with green leaves and other organic materials during the summer months conserves and maintains the population of native beneficial micro-flora. Under common mechanical practices, collecting and destroying disease-infected and insect-infested plant parts is the prerequisite. Collection and destruction of eggs and early-stage larvae of the vectors. Hand picking the older larvae during the early stages of the crop. The infested shoots by the insect vectors must be collected and destroyed. Usage of yellow sticky traps @ 4-5 traps/acre. • Use light trap @ 1/acre and operate between 6 pm and 10 pm. Is very helpful to control vectors. Install pheromone traps @ 4-5/acre for monitoring adult moth's activity (replace the lures with fresh lures after every 2-3 weeks). Common biological practices that can be accomplished to control the vector population are conserving natural enemies through ecological engineering and the augmentation of the release of natural enemies.

15. Cardamom mosaic virus (CDMV)

Etiology: Caused by Cardamom mosaic virus (CDMV) belonging to genus Maculovirus, under family Potyviridae.

Symptom

General chlorosis of young leaves occurs with the formation of the parallel streaks of pale green tissues running along the veins from the midrib to the margins of the leaves is observed. Stich stripes are seen on the leaf sheath. In the progressive stage, the whole plant shows the mosaic symptom. The rhizome shrivels and plants die off. If the young clumps are affected, plants die before flowering (Venugopal, 2002).

Epidemiology

The disease spreads mainly through infected planting materials. It is transmitted through aphids (*Pentalonia caladii*) in a non-persistent manner and infected rhizomes.

Management

Under cultural control, collection and removal of infected clumps along with rhizomes and then burning proved to be effective (Baju and Bhat., 2012). Raising the nursery in disease-free areas is a must. Spraying with dimethoate (or) Methyldematan (or) Phosphomidon to kill the aphid vector is an effective way (Tiwari *et al.*, 2016). Dhanapal *et al.*, (2009) depicted certain effective common management practices for cardamom to avoid all kinds of viral diseases. Almost all cardamom cultivars are susceptible to viral diseases and it is very difficult to control all virus-affected plants by chemical measures along with its deleterious effect on the environment. The important steps to control the spread of viral diseases are undertaking a regular survey in plantations to trace virus-affected plants, Rogueing the affected plants and removing them as soon as they are traced, and repeating the tracing and removal of affected plants in three months' time and interval, proclaiming healthy disease-free seedlings for new planting, Avoiding rhizome planting using clumps taken from virus affected plantations, avoiding transportation of infected materials, avoiding raising nursery near diseased areas., destructing of alternate and collateral host plants of the Zingiberaceae family, planting of virus resistant or genetically transformed plants and finally control vector through spraying of suitable insecticides, bio-insecticides and botanicals.

References

Ajay D., Francis S. M., Vijayan A. K. and Dhanapal K. (2013). Epidemics of Leaf blotch. *International Journal of Current Research*, 5 (8): pp.2109-2112.

Anonymous (2014a). AESA based IPM – Small cardamom, Department of Agriculture and Cooperation Ministry of Agriculture Government of India. https://niphm.gov.in/IPMPackages/Smallcardamom.pdf.

Anonymous (2014b) AESA BASED IPM PACKAGE LARGE CARDAMOM. Department of Agriculture and Cooperation Ministry of Agriculture Government of India pp:1-42

Anonymous (2019). Status of Viral Diseases of Large Cardamom (*Amomum subulatum* Roxb.) and its Management in Sikkim and Darjeeling, West Bengal. Bioinformatics Sub-Disc, SSC&T. http://www.bioinformaticssikkim.gov.in/ Crop%20Diseases/LARGE%20CARDAMOM%202.html.

Anonymous (2021). Agriculture> spice >cardamom > crop management. KAU-Agri-Infotech Portel elkau.in/crops/spices/Cardamom/plant_protection_mainfield. aspx.

Baju C.N and Bhat A.I (2012). Viral diseases of cardamom. Technical buttelin. Pp: 1-13 IISR, G.K printers Azad building Kochi- 17 Cardamom Research Centre Appangala, Madikeri http://iisr.agropedias.iitk.ac.in/sites/default/files/ Viral%20Diseases%20of%20Cardamom.pdf.

Belbase S, Paudel J, Bhusal R, Gautam S, Aryal A and Kumar. S (2018). Fungal Diseases of Large Cardamom (*Amomum subulatum* Roxb.) and Its Integrated Management. *Int.J.Curr.Microbiol.App.Sci.* 7(03): 3316-3321.

Bhat, A.I., Pamitha, N.S., Naveen, K.P. (2020). Identification and characterization of cardamom vein clearing virus, a novel aphid-transmitted nucleorhabdovirus. *Eur J Plant Pathol* 156, 1053–1062.

Dhanapal K., Thomas J., Bhat S. and. Thomas J. (2009). Viral diseases of cardamom and their management. *Spice India,* Vol. XXII (2): pp- 37-39.

Karkee A., and Mandal D. (2020). Efficacy of Fungicides Against *Rhizoctonia solani* Inciting Rhizome Rot Diseases on Large Cardamom (*Amomum subulatum* Roxb.). *International Journal of Applied Sciences and Biotechnology,* 8(1): 61-64.

Mandal B., Mandal S. , Pun K.B. and Varma A. (2007). First Report of the Association of a Nanovirus with Foorkey Disease of Large Cardamom in India, *Plant Disease,* 88(4):428.

Mandal B., Vijayanandraj S., Shilpi S., Pun K.B., Singh V., Pant R.P., Jain R.K., Varadarasan S., Varma A. (2012). Disease distribution and characterisation of a new macluravirus associated with chirke disease of large cardamom. *Ann. Appl. Biol.* 160:225–236.

Nair K. P. (2020). The Geography of Cardamom (*Elettaria cardamomum* M). *Cardamom Pathology*, pp 125-163.

Pathak A. (2021) Facets of the North-east Cultivation of Large Cardamom in Sikkim http://www.kiran.nic.in/pdf/publications/Cultivation_of_large_cardamom_in_Sikkim.pdf.

Paudel J., Belbase S., Gautam S, Bhusal R. and Kumar S. (2018). The Effect of Viral Diseases of Large Cardamom (*Amomum subulatum* Roxb.) on Production and their Management. *Int.J.Curr.Microbiol.App.Sci.* 7(03): 855-860.

Saju K. A, Deka T. N., Gupta U., Bisuos A. K, Vijagan A. K., Thomasz J and. Sudharshan M. R (2010). An epiphytotic of Colletotrichum blight affecting largecardamom in Sikkim and Darjeeling, *Journal of Hill Research*, 23 (1&2): 14-21

Saju A K, Deka T. N., Sudharshan M. R., Gupta U. and Biswas A. K. (2011). Incidence of Phoma leaf spot disease of large cardamom (*Amomum subulatum* Roxb.) and in vitro evaluation of fungicides against the pathogen. *Journal of Spices and Aromatic Crops,* 20 **(2)**: 86–88.

Satyagopal, K., S.N. Sushil, P. Jeyakumar, G. Shankar, O.P. Sharma, D.R. Boina, S.K. Sain, M.N. Reddy, N.S. Rao, B.S. Sunanda, Ram Asre, K.S. Kapoor, Sanjay Arya, Subhash Kumar, C.S. Patni, T.K. Jacob, Santhosh J. E., C.N. Biju, K. Dhanapal, H. Ravindra, Linga Raju, S., Ramesh Babu, B.C. Hanumanthaswamy. 2014. AESA based IPM package for Small cardamom. pp 43.

Thomas J and Bhai S. 2002. Diseases of Cardamom (Fungal, Bacterial and Nematode diseases) In: The Genus Elettaria. (Eds.) P.N. Ravindran and K.J. Madhusoodhanan. Taylor and Francis. 160-179.

Talukdar D. (2021). Chirkey and Foorkey: The dangerous viral diseases of Large cardamom in Sikkim. *Sikkim express* (e-paper), Vol XLV (201):5.

Talukdar D and Anand Y. R. (2021). An insight to Some Important Diseases of Large Cardamom in Sikkim and its Organic management Strategies", under the main book – Plant Health Management- An Insight of Conventional and Modern Approaches", ISBN No- 978-93-5473-043-6, Immortal Publications.

Tiwari S., Anitha P., Shamprasad P. (2016). "Serological Detection of Cardamom Mosaic Virus Infecting Small Cardamom, Elettaria cardamomum L". *International Journal of Life-Sciences Scientific Research*. 2 (4): 333-338.

Venugopal, M.N. (1995). Viral diseases of cardamom (*Elettaria cardamomum* Maton) and their management. *Journal of Spices & Aromatic Crops*, 4 (1): 32-39.

Venugopal, M.N. (2002). Viral diseases of cardamom. In P.N.Ravindran & K.J. Madhusoodanan (Eds.) cardamom-the genus *Elettaria* (pp.143–159). New York: Taylor and Francis.

Vijayan A.K., Thomas J. and Thomas J. (2009). Fusarium diseases, a threat to small cardamom plantations. Spice India, Niseema Printers & Publishers, Kochi-18. Pp-11-13

Vijayan A.K., Saju K.A., Dhanapal K., Eswaran V.M., Vallath A., Pandithurai G., Manoj O, Divya P.V and Manesh K. (2019). Pests and Diseases of Large Cardamom in India and Their Management Practices Under Organic Cultivation. International Journal of Agriculture Sciences, ISSN: 0975-3710 & E-ISSN: 0975-9107, Volume 11, Issue 15, pp.- 8876-8880

Vijayanandraj, S., Mandal, B. Jebasingh T., Jeeva M. L., Makeshkumar, T., Maheshwari Y. (2017). Characterisation of the Macluraviruses Occurring in India. *A Century of Plant Virology in India*, Springer Singapore, pp. 307–326.

Chapter - 14

Diseases of Sugarcane (*Saccharum officinarum* L.) and their Integrated Management

Chandramani Raj[1], Shweta Singh[1], Sanjay Kr. Goswami[1], Matber Singh[2], R. Gopi[3], Chandan Kapoor[4], E. Yogesh Thorat[1]

[1]Division of Crop Protection, ICAR-Indian Institute of Sugarcane Research, Lucknow-226002
[2]Division of HRD&SS, ICAR-Indian Institute of Soil & Water Conservation, Dehradun- 248001
[3]ICAR-Sugarcane Breeding Institute, Research Centre, Kannur, Kerala- 670002
[4]Division of Genetics, ICAR- Indian Agricultural Research Institute, New Delhi-110012

Sugarcane (*Saccharum officinarum*) is an important cash crop from the family Gramineae (Poaceae), class monocotyledons and order glumaceae, subfamily panicoidea, tribe Andripogoneae and sub-tribe saccharininea. Sugarcane finds an important place in the history of Indian culture dating back to Vedic times till today being grown widely. The crop had its origin in New Guinea and has spread along human migration routes to Asia and the Indian subcontinent. It was cross-bred in India with some wild sugarcane relatives to produce the commercial sugar cane known and utilized today. The cultivated canes of today are from two main groups: (a) thin, hardy north Indian types *S. barberi* and *S. sinense* and (b) thick, juicy noble canes *Saccharum officinarum*.

India is the world's largest consumer and the second largest producer of sugar after Brazil. About 2.8 lakh farmers are cultivating sugarcane in an area of 4.4 lakh acres with an annual production of 170 million tonnes and productivity being 78 t/ha. More than 11 crore people in the country are directly or indirectly dependent on this industry in India. It provides employment to over a million people directly or indirectly besides contributing significantly to the national funds. However, the higher production of sugarcane is constrained by the incidence of several diseases and pests. The important diseases of sugarcane causing economic losses have been listed below and discussed in detail in the chapter.

S.No.	Diseases	Causal organism
A.	Fungal diseases	
1.	Red rot	*Colletotrichum falcatum*
2.	Smut	*Ustilago scitaminea*
3.	Wilt	*Fusarium moniliforme*
4.	Rust	*Puccinia kuehnii, P. erianthi* and *P. melanocephala*
5	Pineapple disease	*Ceratocystis paradoxa*
6	Pookah Boeng	*Fusarium fujikuroi*
7	Downy mildew	*Peronosclerospora sacchari, P. philippinensis, P. miscanthi, P. spontanea*
B.	Bacterial Diseases	
1.	Leaf scald	*Xanthomonas albilineans*
2.	Ratoon stunt	*Clavibacter xyli* pv. *xyli*
3.	Red stripe	*Pseudomonas rubrilineans*
C.	Viral Diseases	
1.	Grassy shoot	*Candidatus Phytoplasma sacchari*
2.	Yellow leaf	Sugarcane Yellow Leaf Virus
3.	Nematode	

1. Red rot

Economic importance

Red rot is the most devastating sugarcane disease. The disease is a major limiting factor in sugarcane yield (Viswanathan and Rao 2011) and is also known as the cancer of sugarcane (Sharma and Tamta 2015). The red rot disease is responsible for wiping out elite sugarcane varieties (Viswanatahan and Rao 2011; Sharma and Tamta 2015). The disease reduces sugar by 75% in the field (Viswanathan and Samiyappan 1999; Hussnain and Afghan 2006; Viswanathan and Rao 2011).

Red rot caused by *Colletotrichum falcatum* Went causes 5–50% losses in sugarcane production, which further decrease the sugar recovery by 31%. Red rot is the most widely spread disease in India and sugarcane growing regions of the world and is reported to kill sugarcane plants (Hossain 2020).

Symptoms

The disease is destructive in India due to the prevalence of high humidity and temperature. Drying and discolouration of the stalk and spindle leaves, the presence of white spots in internodal tissues reddening of internodal tissues and the smell of vinegar are the main symptoms of the diseases for quick identification. The stalk is more vulnerable to this disease and it is called stalk and seed plant disease (Suman *et al.*, 2005). The pathogen invades the leaf blade and leaf-sheath including symptom appearance on the midrib of the leaf and stalk. According to Duttamajumder 2008, disease signs appear as the death of young and emerging shoots in March-May in north India. Cottony grey fungal mass develops in the pith region of the internodes. Nodal rotting appears when the crop is at the end of the growth phase during August-September in sub-tropical India. In the early stages of infection, it is difficult to recognize the presence of the disease in the field, as the plant does not display any external symptoms or distress. Some discolouration of rind often becomes apparent when internal tissues have been badly damaged and are rotten. Sometimes tiny red spots on the upper surface of the midrib are visible (Sharma *et al.*, 2017). The death of a few plants or clumps leads to the failure of the entire crop (Duttamajumder, 2008). The disease-affected plant may die within 10–15 days (Saksena *et al.*, 2013).

Causal organism

Red rot is incited by the fungus *Colletotrichum falcatum* Went (Teleomorph: *Glomerella tucumanensis* [Speg.] Arxand Muller). The pathogen produces both imperfect and perfect stages. According to Abbot (1938), there are two races of isolates of *C. falcatum* i.e., dark and light races. The mycelium of the fungus in host tissue is mostly intracellular and grows mainly in the parenchymatous cells of the pith. The hyphae are thin branched, hyaline, and septate. A large number of black setae develop in and around the stroma. All strains of *C. falcatum* do not produce setae.

The sickle-shaped or falcate conidia measure 16 to 40 μm x 5 to 7μm in size and contain oil globules in the middle. Mostly the conidia are produced in the acervuli but sometimes produced directly on the mycelium in the culture medium or on the infected stalk. Many strains of the red rot organism in old culture and diseased stalk produce round, double-walled structures called chlamydospores. These remain dormant in the soil for varying periods. *C. falcatum* is a facultative parasite and it keeps on mutating.

Rafay and Singh (1957) divided the *C. falcatum* strains into 9 types *viz.*, Type A, Type B, Type C, Type D, Type E, Type F, Type G, Type H and Type I. In India the perfect stage was first announced by Chona and Srivastava (1952), they observed it on the culture medium. The ascus contains 8 hyaline ascospores, straight to slightly fusoid, 1-celled and measures 18-22x7-8 μm.

Disease cycle

The red rot infected setts/soil have conidia and chlamydospores. These conidia may germinate in the soil and form appressoria which act like chlamydospores. Diseased sugarcane produced conidia in the rind of the cane, midrib and leaf blade surface. Conidia in the soil are spread by irrigation water, air and rain which causes secondary infection to healthy canes.

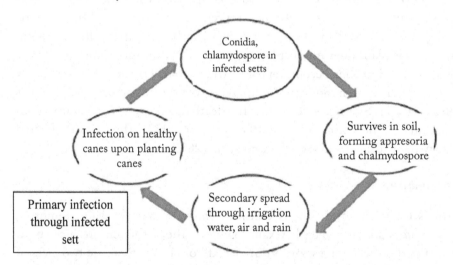

Epidemiology

C. falcatum infects mature sugarcane, and leaf midribs and causes rotting of the setts, which causes significant losses in sugar quality and crop yield (Rao 2004). The pathogen is sett-borne. The perpetuation of the disease is through infected setts, canes, diseased stubbles, debris and resting propagules in the soil. The recurrence of red rot in sugarcane is primarily due to infected setts and stubbles which take the pathogen to the next crop [Butler and Khan 2013; Agnihotri *et al.*, 1979). The pathogen is not a true soil-borne fungus and survives in the soil for 5-6 months only (Chona and Nariani 1954; Singh *et al.*, 1977).

Integrated management

a. Prophylaxis

Prophylaxis is an important method to manage diseases in plants. In this approach, keep the pathogen away from the host to reduce or avoid the inoculums of the pathogen. Phytosanitary or field sanitation measures are adopted i.e., removal and destruction of debris, sett cutting knives should be disinfected. Proper drainage is essential in the field. The crop rotation can reduce the load of the inoculum. The sett showing reddening at the cut ends or the nodal region should be discarded. Ratooning should avoid. The diseased crops should be harvested as early as possible to avoid losses and disease spread.

b. Resistant varieties

Resistant or moderately resistant varieties should be planted.

c. Heat treatment

Moist hot air treatment (MHAT) at 54 °C for 4 h is effective against red rot (Singh 1977).

d. Chemical treatment

Studies on the management of red rot with fungicides showed that fungicide chemicals protected the planted setts from soil-borne inoculum and improved germination. Foliar spray of chemical fungicides is not effective in the management of the disease. However, better crop stands have been achieved from increased germination by treating setts with fungicides before planting. The organo mercurials like Aretan, and Agallol (0.25%) for a 5-10 minute dip help in the eradication of superficial inoculum but not the deep-seated mycelium. Antifungal compounds like thiophanate methyl are specific against *C. falcatum* (Satyavir, 2003) along with Benomyl 50 WP, Ridomil 75 WP and Folicar which are effective in controlling the pathogen (Subhani *et al.*, 2008). The application of these fungicides reduces the incidence of red rot infection in the seed. Sugarcane synthesizes phytoalexins as luteolinidin and apigeninidin when inoculated with *C. falcatum*, revealing the possible role of these compounds in red rot resistance (Viswanathan *et al.*, 1994). The treatment of sett with *Trichoderma* spp., *Pseudomonas* spp., *Bacillus* spp. and timely application of chemicals manage the red rot effectively (Li *et al.*, 2020).

2. Smut

Economic importance

Smut is an important disease of Sugarcane across the growing areas of the crop. This disease was first observed in 1877 in Natal, South Africa and reported to be due to a fungus *Ustilago scitmainea* H. & P. Sydow in 1924. The fungus was renamed *Sporisoriun scitaminea* (Syd.) M. Piepenbr., M. Stoll & Oberw in 2002. Till the 1950s, this disease was of concern only in Asia; however, in the present condition, it has spread to sugarcane-producing areas of many countries like *viz.*, Indonesia, Morocco, Iran, Rhodesia, Mauritius, the islands of Java, Sulawesi and Sumbawa, Argentina, Hawaii, the Caribbean, the mainland USA, Central America and Southern Brazil. In India, smut disease became a severe problem to sugarcane production during the 1930s. The sugarcane production in Maharashtra and Northern Karnataka suffered significant yield loss due to this disease. Although the disease is present all over India, severe damage occurred in tropical regions *viz.*, Andhra Pradesh, Maharashtra, Tamil Nadu, Karnataka and Gujarat of the country. Cane yield losses of 39-56% have been reported in planted crops whereas 52-73% yield loss was noticed in the ratoon crop.

Symptoms

The emergence of elongated whip-like sorus in the apical region of the cane is the most characteristic symptom of this disease. The size and shape of whips vary according to the variety and favourable conditions. It may be short (a few centimetres) to long (up to 1.5 m), twisted, single or multiple whips on the crop canopy. The emerged whips are mainly composed of the central core of host tissue enveloped by a thin layer of black spores under a thin silver-white membrane. The affected canes often tiller profusely with small narrow leaves and more erect shoots which gives a bushy or grassy appearance. Leaf and stem galls or bud proliferation can also be seen in this disease.

Causal organism

The smut disease of sugarcane is caused by the fungus *Ustilago scitaminea* H. & P. Sydow which was renamed *Sporisorium scitamineum* (Syd.) M. Piepenbr., M. Stoll & Oberw. 2002. The fungus belongs to the order Ustilaginales under Basidiomycota.

Disease cycle

The primary infection occurs through infected setts from smutted canes which do not show whip symptoms and are used for seed purposes. The terminal or axilliary whip serves as a source of inoculum in the field. Teliospores are released from these whips and infect the healthy buds which may remain dormant or exhibit the symptoms. Spores present on the soil surface may enter the neighbouring field through irrigation water or splashed rain.

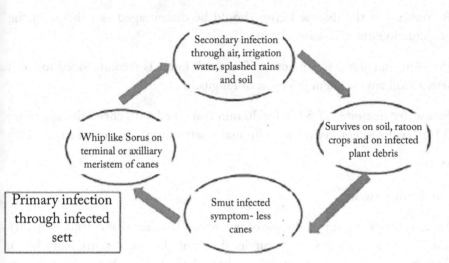

Epidemiology

Hot and dry weather conditions are suitable for smut infection and spread. Ratooning of infected canes and any stresses on the sugarcane plants increases the frequency of whip development. Moreover, high rainfall and severe winter reduce the severity of smut development.

Integrated disease management

a. Avoid planting infected setts from smutted canes as it invariably leads to smut infection in field.

b. Seed canes should be collected from disease-free regions to avoid mixing of any smut spores from diseased to healthy canes. Fields with more than 2% disease incidence should not be selected for seed purposes.

c. Seed treatment with recommended disinfectant should be carried out before planting.

d. Practice of rouging of smutted canes should be followed upon first observation of whip-like sorus in the field. This prevents the spread of the disease in the same season as well as reduces the amount of carryover for the next season. In the process of rouging, smutted shoots are first covered with a gunny bag or sack or polythene and then cut from the base. The collected smutted shoots should be either burnt or buried in the ground.

e. Ratooning of the diseased crop should be discouraged as it helps in the perpetuation of the disease.

f. Pre-plant hot water treatment at 50°C for 2 hours is recommended for seed setts to kill any incipient infection of fungus.

g. Hot water treatment @ 52°C for 30 min combined with chemotherapy using 0.1% Triademiphon completely eliminates sett-borne infection of smut.

3. Wilt

Economic importance

Wilt is an important complex disease of sugarcane. The disease incidence may vary from 0-75% in sugarcane varieties under different climatic conditions in India. Andhra Pradesh, Orissa, Gujarat, Bihar, Uttar Pradesh, Punjab and Haryana are the most affected states, while in Maharashtra, Tamil Nadu, Kerala and Karnataka the disease incidence has been less.

Earlier, the disease was known to occur in Bihar (1940-1945), but now the disease spread in many states like Punjab, Uttar Pradesh and Haryana. In Tamil Nadu during 1955-1956, wilt caused considerable loss to the varieties Co 419, Co 449, Co 453, Co 527, Co 1122 and H32-8560 (Srinivasan, 1964) where incidence ranged from 5-80%. Sugarcane varieties namely Co 527, Co 951, Co 1007, Co 1223, CoS 245 and CoS 312 were discarded due to severe wilt problems (Kirtikar *et al.*, 1972). In Andhra Pradesh, wilt causes severe loss in the sugarcane cultivars like Co 419, Co 527, Co 775, Co 975 and Co 997 during 1959. The disease incidence occurs in all the major sugarcane-growing areas of the country (Viswanathan and Rao 2011; Viswanathan 2013). Any field under harvest displays 10-15% of dried canes and 50% of them were affected due to wilt (Viswanathan 2018).

Symptoms

Yellowing of foliage, development of cavity in the internodes, browning of vascular tissues and blackening of roots and reduction in its volume are the main symptoms of the disease.

All the stages of the sugarcane are affected by the disease *viz.*, germinating, young and mature crops. The mid-rib of leaves becomes yellow, and the plants become stunted. The disease symptoms are known to appear in the monsoon or post-monsoon period. At the maturity of the crop, the canes become light and hollow. In early infection, split canes show the purple or muddy red colour as conical patches at each node. The cane emits a bad odour which is different from the red rot sour smell. The spindle-shaped cavities develop in each internode due to the desiccation of tissues. The cavities may also develop in nodal tissue. The walls of the xylem vessels become dark and hyphae remain in the lumen.

Causal organism

Fusarium spp., *Cephalosporium* spp. and *Acremonium* spp. are supposed to be the incitants of the sugarcane wilt. The *Fusarium* isolates produce microconidia and macroconidia. The macroconidia are straight to falcate and the apical and basal cells were either conical or blunt. The associated pathogen has not been established with the disease. The studies conducted by ICAR-IISR, Lucknow showed that the wilt of sugarcane is caused by *Fusarium sacchari*.

Disease cycle

The wilt fungus overwinters in crop debris and sett. The perithecia are formed on the surface of the stalk, which releases ascospores spreading through wind and rain. The infection occurs through root and stalk wound. This leads to the wilt disease of sugarcane, which shows pink to reddish brownish discoloration of pith.

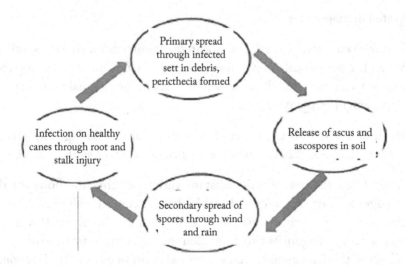

Epidemiology

The disease occurs under three conditions *viz.*, virulent pathogen, favourable environment and susceptible host. Pathogen-infected seed, crop residues and the pathogen in the soil serve as the primary source of infection. It is found that *F. sacchari* survives in the soil for 2.5–3 years (Viswanathan, 2012).

Disease outbreak depends on the environmental factors which influence the disease severity. The secondary spread of the disease from one place to another is caused by rain and irrigation water. In this disease primary source of inoculum is the soil-borne inoculum which is critical in plant infection. Infected setts readily initiate infection on the root and disease progresses to the growing sprout. Combined infections of *F. sacchari* and *C. falcatum* have also been observed (Viswanathan *et al.*, 2006). The damage is more when wilt and red rot occur together (Agnihotri and Rao, 2002). *C. falcatum* weakens sugarcane and facilitates the infection of *F. sacchari*. The infestation of root borer, (*Polyacha depressella*) predisposed the plant to wilt. Sardana *et al.* (2000) showed a positive relationship between root borer and wilt in the cultivar Co 89003 in the subtropical region and in another cultivar TNAU Si8 (Si 2000–02) in Tamil Nadu (Viswanathan, 2013).

In wilt epidemic studies in the National Hybridization Garden at ICAR-SBI, Coimbatore, clones expressed wilt either alone or in combination with different insects. Viswanathan *et al.* (2015) reported that moisture deficiency in summer, the high day temperature and low humidity also increased wilt incidence. Recently, sugarcane wilt has been reviewed in detail by Goswami *et al.*, (2020).

Integrated management

a.   Management of wilt is a serious issue in the country as the disease is soil and sett borne. The use of healthy and disease-free seeds, field sanitation, crop rotation, resistant varieties, and chemical seed treatment are the main practices for the management of the wilt.

b.   The use of wilt-resistant varieties is economical and ecologically sustainable. Viswanathan *et al.* (2019) found resistance in 22.3% of sugarcane accessions.

c.   Apart from the use of wilt-resistant varieties, other methods for disease management are also required like biological control, cultural methods and chemical methods, which can be integrated to manage the disease. Potential fungal and bacterial antagonists have been identified against sugarcane pathogens. The efficacy of these antagonists has been proved under *in vitro* and *in vivo* conditions by Viswanathan and Malathi (2019). Bhatti and Chohan (1970) showed the wilt suppression by *Bacillus* and *Streptomyces* antagonists. The antagonistic efficacy of *Trichoderma* against wilt in endemic areas is observed by Viswanathan *et al.* (2012). *T. harzianum* isolates Th1 and Th2 showed hyperparasitism on *F. sacchari* (Viswanathan *et al.*, 2014).

d. Healthy seed canes are very important for disease management. This exercise prevents the entry of the pathogen into new areas. Disease severity increases in ratoon crops. So, wilted plants should be harvested early and ratooning should be avoided. After harvest, the residue should be burned (Joshi, 1954). This practice reduces the inoculum level of the pathogen. To decrease the soil-borne inocula of the pathogen, crop rotation with paddy is followed in disease-endemic regions.

e. The plant becomes susceptible to wilt when predisposed to waterlogging, drought and borer pests. Proper care should be taken to avoid abiotic stresses.

f. Sett treatment with carbendazim (0.1%) is effective in controlling wilt diseases. Viswanathan and Padmanaban (2008) showed that the use of Chloropyriphos 20EC/ Imidacloprid @ 0.5L a.i./ha/Carbofuran 3G/ Fipronil @1.5 kg a.i./ha during the last week of August effectively managed the borer and wilt severity in the cv. Co 89003 in subtropical India. Application of 40 ppm boron or manganese to wilt sick soil decreases the wilt incidence. The incidence of the disease reduces when healthy sets are dipped in a similar solution before planting in sick plots (Ganguly and Jha 1964).

g. Hot water treatment (HWT) at 50°C for two hours has been found effective (Srinivasan and Rao 1968).

4. Rust

Economic importance

Rust is an important foliar disease of sugarcane crops which has caused severe economic losses to crop cultivation in more than 60 countries of the world. This disease was first described by Krüger in 1890 who suggested the causal organism as *Uromyces kuehnii*. Later the pathogen was renamed *Uredo kuehnii* by Wakker and Went. The observation of teliospores by Butler (1918) laid the change of genus to *Puccinia* as *P. kuehnii* Butler. Till now, three different types of rust have been reported based on the morphology of spores; 1. Brown rust (caused by *Puccinia melanocephala* (Syd. & P.Syd)); 2. Orange rust (incited by *P. kuehnii* (W.Kruger) E.J. Butler) and 3. Tawny rust (caused by *Macruropyxis fulva* sp.) Of three rusts, brown rust is more prevalent across the growing areas of sugarcane whereas orange rust is reported in Asia, Australia, and Oceania while tawny rust is restricted only to the South African continent. In India, epidemics of brown rust of sugarcane were observed in Maharashtra and Himalayan foothills of Uttar Pradesh on popular variety Co 475 and CoS 510 in 1949 and 1956 respectively. The short life cycle, airborne nature of the pathogen and continuous presence of susceptible varieties as main or ratoon crop over large areas has resulted in a severe outbreak of the disease which eventually leads to the withdrawal of many popular varieties from cultivation. The economic yield loss has been reported from 10-50% under different conditions in many countries.

Details of outbreaks and epidemics of rust of sugarcane across the cultivation areas

Place	Country	Variety	Year	Pathogen
Africa South of Sahara	South Africa	Co 301	1941	*Puccinia kuehnii* now confirmed as *P. melanocephala*
Angola and Cameroon	Africa	B 4362	1978	*Puccinia kuehnii, Puccinia erianthi*
Matanzas province	Cuba	Ja60-5	1998	*Puccinia melanocephala*
Southern Florida	USA	CP78-1247	1988	*Puccinia melanocephala*
Maharashtra	India	Co 475	1949	*Puccinia melanocephala*
Uttar Pradesh	India	CoS 510	1956	*Puccinia melanocephala*

Symptoms

Rust disease can be easily identified by the presence of orange to brown powdery spore mass after rubbing the open pustules between the fingers. At the beginning small to elongated yellowish spots or flecks develop on the upper and lower surface of leaves which become brown to orange-brown or red-brown with a yellow halo around the lesion. However, orange rust lesions never turn into dark brown lesions as in brown rust. These lesions develop into pustules after the formation of uredinium on the underside of the leaves. Later, chlorotic lesions turn necrotic when orange pustules become black. In severe infection, the individual leaves appear brown or rusty due to the collision of several necrotic lesions with each other which results in the death of the leaves before maturity that giving a burnt appearance distance.

Causal organism

Rust of sugarcane is caused by two different species of *Puccinia, Puccinia melanocephala*

(Syd. & P.Syd) and *P. kuehnii* (W.Kruger) E.J. Butler and *Macruropyxis fulva* sp. nov. The difference between species can be determined through reaction on differential varieties.

Details of difference between common rust and yellow rust pathogen

Character	*P. kuehnii*	*P. melanocephala*
Urediniospores	Larger, paler with thick walls on the apical portion	Smaller, darker coloured spore walls, prominent pores and no apical thickening
Urediniospore ornamentation	spines are 3 – 4 μm apart and not clustered over pores	spines are 1 - 1.5μm apart and clusters over pores
Paraphyses	few	abundant
Teliospores	rare	readily found

Disease cycle

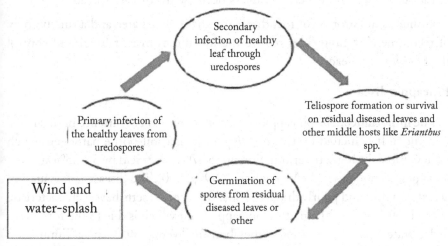

Epidemiology

The major epidemiological factor for rust development is the age of the plant, temperature and humidity. Two and six-month old younger plants have shown more susceptibility than old age. The pathogen grows luxuriously in cool weather as compared to hot conditions and causes significant disease at a temperature of average 18-26°C. The free moisture of about 65-70 hours on the leaf surface at this temperature favours spore germination, infection and subsequent spread of the disease. The outbreak of rust usually occurs due to the continuous presence of sugarcane in the field as the main crop or ratoon, growing a single susceptible variety over a larger area and conducive environmental factors for the pathogens.

Integrated disease management

a. The selection of resistant varieties and withdrawal of susceptible varieties from cultivation is the most economic and effective approach for the management of rust. If the susceptible varieties are the popular ones and farmers do not have any other choice then they can be allowed to grow only in localities which are unfavourable to rust pathogen.

b. The scheduling of irrigation and fertilization (rational application) of the crop at a suitable time reduces rust disease.

c. The application of organic fertilizer to the crop has been reported to increase the immunity of the crop against rust disease.

d. Phytosanitary measures like removal of diseased leaves, stripping off old leaves, discarding the weak plants and weeding at timely interval enhance the ventilation of the sugarcane field which eventually decrease the disease spread.

e. Continuous monitoring of rust disease in the infected area and a timely spray of recommended fungicides reduce the inoculums of rust fungi and control their epidemic spread.

5. Pineapple disease

Economic importance: The pineapple disease is of common occurrence in sugarcane fields in India. It is incited by *Ceratocystis paradoxa*, a soil-borne pathogen which causes major loss during sett germination up to 47%; cane yield by 31-35% and 10-15 tonne per hectare in yield (Anonymous, 1999; 2000). The disease is favoured in heavy textured soils and poorly drained fields. The affected setts have a characteristic odour resembling that of the mature pineapple fruit which is due to the formation of ethyl acetate during the pathogen metabolism (Went, 1896). The accumulation of ethyl acetate in the tissues may rise up to 1% which has the capacity to inhibit the germination of buds (Kuo *et al.*, 1969).

Causal organism: *Ceratocystis paradoxa*

The pathogen produces both macroconidia and microconidia. Conidiophores are linear, thin-walled with short cells at the base and long terminal cells. The microconidia are hyaline when young but become almost black at maturity. They are thin-walled, cylindrical and produced endogenously in chains in the long cells of conidiophores and pushed out in succession. Macroconidia are produced singly or in chains on short, lateral conidiophores. Macroconidia are spherical or elliptical or truncate or pyriform and are hyaline to olive green or black measuring 16-19 x 10-12 μm. The fungus also produces chlamydospores on short lateral hyphae in chains, which are oval, thick-walled and brown in colour. The perithecia are flask-shaped with a very long neck. The bulbous base of the perithecium is a hyaline or pale yellow, 200-300 μm in diameter and ornamented with irregularly shaped, knobbed appendages. The ostiole is covered by numerous pale-brown, erect tapering hyphae. Asci are clavate and measure 25 x 10 μm and ascospores are single-celled, hyaline, ellipsoid, more convex on one side, and measure 7-10 x 2.5-4 μm.

Symptoms: The disease primarily affects the setts usually two to three weeks after planting. The fungus is soil-borne and enters through cut ends and proliferates rapidly in the parenchymatous tissues destroying the central soft portion i.e. parenchymatous tissues of the internode and then damaging the buds. The affected tissues first develop a reddish colour which turns to brownish black in the later stages. The severely affected setts show internodal cavities covered with mycelium and abundant spores. A characteristic pineapple smell is associated with the rotting tissues upon cutting open the diseased stalk. The setts may decay before the buds germinate or the shoots may die after reaching a height of about 6-12 inches as the presence of fungus inside the set prevents rooting. Infected shoots are stunted. Occasionally, the disease occurs in standing crops too due to the entry of the pathogen through stalks damaged by borers, rats or any such injuries.

Disease cycle

The disease spreads primarily through micro-conidia/chlamydospores present in the soil. The secondary infection is through blown wind, irrigation water and rain-splashed spores gaining entry through damaged tissues due to mechanical and animal injuries. The insects like cane borer (*Diatraea dyari*) also help in the spread of the disease.

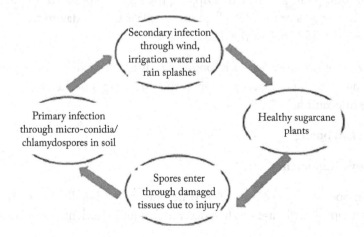

Epidemiology

The disease is favoured in poorly drained fields with heavy clay soils where excessively deep planting is done. The lower temperature condition of soils and prolonged rainfall after planting also increases the disease incidence (Moutia and Saumtally, 1999; Man, 2007).

Integrated disease management

a. The easiest and most economic method to manage the disease is the use of a resistant cultivar. Disease-free setts should be used since the major source of resistance is in the infected plants and setts.

b. The farmers usually treat setts with 2-3% lime solution for 12-24 hrs to eliminate the pathogen before growing.

c. The cut sides of setts should be covered with lime or in combination with $CuSO_4$ solution to avoid infection (Man 2007)

d. Field surveys are recommended to detect, remove and destroy completely the infected plants and lime should be applied to the infected places prior to replacement with disease-free setts to reduce disease inoculums in the field.

e. Crop rotation between the cane crops with soybean and/or peanut for several years may also interrupt the disease cycle and reduce its impact on plants in the field.

f. Dry and well-drained soils for planting cane are recommended for farmers.

g. Several types of fungicides such as Propiconazole and Carbendazim can be used to inhibit pathogen growth. Dipping the setts in 40ppm Boron or Manganese for 10 minutes or in 0.25% Emisan or 0.05% Carbendazim for 15 minutes has proved helpful against the disease.

h. Avoiding the practice of ratooning in diseased fields and growing coriander or mustard as a companion crop in the early stages of the crop will also reduce the infestation.

6. Pookah boeng

Economic importance

"Pokkah boeng" is a Javanese term which describes this disease as "malformed or distorted top". The disease was first described in Java by scientist Walker and Went in

1896 and later reported by Edgerton (1955) and Martin *et al.* (1961). With the advent and spread of the popular variety POJ 2878 during the 1920s, this disease flared into an epidemic in Java. Till the 1970s, the disease usually occurred in a sporadic form in other parts of the world and caused no threat to sugarcane production. However, a severe outbreak of pokkah boeng disease occurred in Guangxi, China during the 1980s. Recently, the disease has spread to all the countries wherever the crop is grown in a commercial manner. In India, the disease was first noticed during the 1980s in Maharashtra on varieties Co 7219 and CoC 671 by Patil and Hapase and later the incidence was also observed in another part of the country. Generally, the disease has been reported to cause 10-38% yield losses in the susceptible variety of sugarcane. However, the pathogen has the potential to cause damage up to 90% in favourable conditions. Besides yield reduction of crops, it also affects the quality of products due to the secretion of several toxins and enzymes by the pathogen. Depending on cultivars, sugar recovery has also been reported to be hampered by 40.8-64.5% due to this disease. Although the disease was considered of minor importance earlier, the recent past witnessed a major threat to sugarcane cultivation in all cane-growing areas of the country.

Symptoms

The symptom of "pokkah boeng" disease was first described by Dillewijn in 1950 as "malformed and twisted top". The early symptom appears as chlorotic areas at the base of young leaves and occasionally on the other parts of the leaf blades. Often the base of affected young leaves becomes narrower compared to normal leaves. Wrinkling, twisting or tangling and splitting of spindle leaves, ladder-like lesions, red stripes, shortening and malformation or distortion of the young leaves are the prominent symptoms of pokkah boeng disease. In the highly susceptible varieties, leaf infection further enters into the stalk through its growing point and in the end, the heavy infection kills the entire top. The symptoms also appear on leaf sheaths as chlorosis and asymmetrical necrosis areas of a reddish colour. The reddish tissue often forms ladder-like lesions which sometimes break through the surface of the rind.

Causal organism

The Pokkah Boeng is caused by a pathogen named *Fusarium fujikuroi* (formerly *Gibberella fujikuroi*) species complex (FFSC). A total of eleven different species of *Fusarium* have been reported to be associated with this disease around the world (table 1). The pathogen belongs to Division: Ascomycota, Class: Sordariomycetes, Order: Hypocreales and Family: Nectriaceae.

Table 1. Detail of *Fusarium* species associated with Pokkah Boeng disease around the world

Country	Causal Organism	Reference
Indonesia	*F. anguioides* Sherb.; *F. bulbigenum* Cke. and Mass. var. *tracheiphilum*; *F. moniliforme* Sheld. var. *subglutinans*; *F. moniliforme* Sheld. var. *anthophilum*; *F. neoceras* Wr. and Rkg.; *F. orthoceras* App. var. *longius* Wr. ; *F. semitectum* B. and Rav	Semangun (1992)
Thailand	*F. moniliforme* Sheldon	Giatgong (1980)
Malaysia	*F. moniliforme* var. *subglutinans*	Geh (1973)
China	*F. verticillioides*	Hilton *et al.* (2017)
S. Africa	*F. andiyazi, F. proliferatum* and *F. Sacchari*	Govender et al. (2010)
India	*F. moniliforme*	Patil and Hapase, (1987)
India	*F. moniliforme* Sheldon	Patil *et al.* (2007)
India	*F. moniliforme* var. *subglutinans*	Vishwakarma *et al.* (2013)
India	*F. sacchari* and *F. proliferatum*	Viswanathan *et al.* (2019)
Java	*F. moniliforme* Sheldon	Bolle (1927)
China	*F. verticillioides*	Lin *et al.* (2014)
China	*F. verticillioides* and *F. proliferatum*	Zhang *et al.* (2018)
Brazil	*F. sacchari, F. proliferatum* and *F. andiyazi*	Costa *et al.* (2019)
Elsewhere	*F. moniliforme* var . *subglutinans*	Matsumoto, 1952
Malaysia	*F. sacchari* not *F. proliferatum* and *F. verticillioides*	Nordahliawate *et al.* (2008)
Egypt	*F. verticillioides*	Osman, 2021

Disease cycle

The primary infection occurs through conidia of pathogen travelled through air currents from infected plant debris to healthy young spindle leaves. The injury made by borers or any other insects or hail damage or natural growth cracks provides access to pathogen into the host tissues. Pathogen proliferates in favourable conditions and causes secondary infection through infected setts, irrigation water, splashed rains and soil. The pathogen can survive for 12 months in the plant debris under natural conditions.

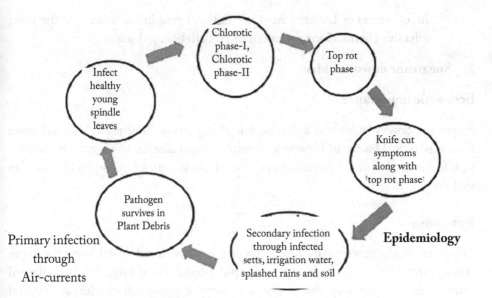

Epidemiology

Pokka boeng diseases generally appear during hot and humid conditions of the environment. The relative humidity (70-80%), and cloudy weather with drizzling rains favour the growth and spread of the pathogen. Sowing of Infected sett, constant cultivation of susceptible variety in the field, exposing sugarcane plants to water stress followed by heavy rain, drenched conditions of the soil, lack of cultural practices with heavy loads of weeds and cultivation of other susceptible varieties in the surroundings are the major epidemiological factors for the development of disease.

Integrated management

» Selection of healthy cane for the planting of sugarcane is highly recommended for this disease.

» The cultivation of resistant varieties is always advisable for t farmers.

» The canes showing 'top rot' or 'knife cut' should be rogued-out immediately.

» The practice of crop rotation should be followed to reduce the inoculum load present in the soil

» Two to three sprays with an interval of 15 days interval of fungicides like Bavistin @0.1% (1 gm/ lit. of water) or Copper oxychloride @0.2% (2gm/

lit. of water) or Dithane M-45 @ 0.3% (3 gm/ lit. of water) are the most effective chemical management of pokkah boeng disease.

7. Sugarcane downy mildew

Economic importance

Sugarcane downy mildew is a disease found in parts of Southeast Asia and some Pacific islands. Downy mildew is a systemic fungal disease of sugarcane that has significant economic consequences in susceptible varieties causing yield losses up to 40%.

Symptoms

The disease is characterized by downy fungal growth with yellow stripes on the upper surface, shredding of older leaves, and rapid elongation of internodes of affected canes. There are two easily identifiable symptoms of downy mildew: leaf streaks and leaf shredding. The two symptoms are related to different stages in the pathogen life cycle. Classic leaf streak symptoms are initially white, turning yellow and then brick red with age. They are associated with the asexual stage of the pathogen and can be seen at any time of the year. Streaks run parallel to the leaf veins. A white downy growth is produced on the underside of affected leaves on hot humid nights. The downy growth is the diagnostic symptom of the disease, and gave rise to the name 'downy mildew'. The leaf shredding symptom is associated with the sexual stage of the pathogen and is more common at cooler times of the year. The pathogen produces a large number of oospores inside the leaf, causing the leaf to shred. Leaf shredding is less common than the leaf streaking symptom.

Causal organism - *Peronosclerospora sacchari, P. philippinensis, P. miscanthi, P. spontanea*

P. philippinensis is an obligate pathogen which requires a living host for it to grow and proliferate. Infection occurs when airborne conidia from an infected crop attach to a susceptible crop host, but seed-borne infections are probably important means of spreading pathogens. A higher rate of infection occurs at a temperature above 16°C (Jepson, 2008). Inside the host, the pathogen produces mycelium which gives rise to conidiophores bearing conidia. The germinating conidia produce germ tubes and penetrate the meristematic tissues or stomata forming haustoria. These haustoria extend throughout the tissue forming mycelium, and the infection continues. The mycelia are branched, slender (8 μm in diameter) and irregularly constricted. The conidiophores are dichotomously branched, measuring 15-26 x 150-400 μm. The conidia are hyaline, ovoid to round cylindrical, slightly rounded at the apex, 17-21 x 27-39 μm. The haustoria are simple, vesical forms to sub-digitate (Weston 1920; Smith and Renfro, 1999). Oospores which are rarely produced are spherical, smooth-walled and approximately 22 μm in diameter.

Disease cycle

The primary source of infection is through oospores in soil and also dormant mycelium present in the infected setts. The secondary spread is through airborne sporangia. Some species are also seed-borne. At the onset of the growing season, at soil temperature >20°C, oospores in the soil germinate in response to root exudates from susceptible seedlings. These pathogens also invade the sugarcane plant via conidia landing on young buds and young leaf tissue at the base of the leaf spindle in young shoots. The pathogen invades the stalk tissue and moves through the cane plant to infect newly-developed leaves. With time, these show the characteristic leaf striping symptoms. This disease can also infect other grasses and crops such as miscanthus, corn and sorghum. If the pathogen is seed-borne, whole plants show symptoms. Oospores are reported to survive in nature for up to 10 years. Once the fungus has colonised host tissue, sporangiophores emerge from stomata and produce sporangia which are wind and rain splashes disseminated and initiate secondary infections. Sporangia are always produced at night. They are fragile and cannot be disseminated more than a few 100 meters and do not remain viable for more than a few hours. Germination of sporangia is dependent on the availability of free water on the leaf surface. Initial symptoms of disease i.e., chlorotic specks and streaks parallel to veins occur within 3 days. As the crop approaches senescence, oospores are produced in large numbers.

Epidemiology

Downy mildew infects the sugarcane plant via conidia infecting lateral buds or the basal portions of the young spindle leaf (Leu & Egan 1989). With the latter, the conidia germinate and the resulting mycelium grows down into the base of the leaves and up towards the growing point. From here, the pathogen infects the plant systemically leading to symptoms in developing leaves. Infected lateral buds may germinate to produce infected side shoots on standing stalks or they can produce systemically infected plants if the buds are planted. Infection from oospores has been reported for *P. miscanthi* (Chu 1965) but the role of oospores in the infection cycle of the other species is not clear and some authors consider that they play no role (Hughes & Robinson 1961; Leu & Egan 1989; Suma & Magarey 2000).

Integrated disease management

Cultural, chemical, and biological control, and HPR are the components of integrated disease management.

a. **Host plant resistance (HPR)** is a practical and economic method of control for downy mildew. This is a unique technology embedded in seed, without direct cost to farmers and does not require extra effort and understanding for its use which requires an effective screening technique; good sources of genetic resistance; a proper breeding method to efficiently incorporate the resistance; a sound strategy for cultivar deployment; and effective monitoring system for pathogen virulence and resistance durability.

b. **Cultural practice:** Cultural methods for controlling downy mildews are largely aimed at sanitation and manipulation of the environment to the advantage of the host and to the detriment of the pathogen. Since the pathogens survive in the form of oospores in the host tissues removal, destruction and burning of the infected plant debris along with weeds serve to reduce the primary inoculum (Butler, 1918; Vasudeva, 1958). Clean, well-drained soils with a two-year crop rotation with a non-host crop have been recommended. Avoidance of monoculture and growing the same variety in particular fields reduces the inoculum buildup and restricts virulence selection in the pathogen population.

c. **Chemical:** The advent of Metalaxyl (methyl N- (2,6-dimethylphenyl)-N- (methoxyacetyl)-dl-alaninate), a systemic fungicide, provided a real breakthrough in the control of downy mildews. The fungicide is absorbed through the leaves, stem and roots and inhibits protein synthesis in the fungus. It has various formulations and can be applied as a seed treatment or foliar spray.

9. Leaf scald

Economic importance

Leaf scald was first recognized as a bacterial disease of sugarcane in the 1920s. The disease has been found in at least 60 countries most of them being major sugarcane countries. Leaf scald is a disease that can lead to sugarcane death. It causes extensive yield losses in highly susceptible varieties through the death of stalks and poor ratooning. Therefore, it can have a great effect on sugarcane yields and has the potential to seriously limit the cultivation of susceptible varieties. The disease is insidious in that it may have a latent (asymptomatic) period that lasts for months and sometimes years. Leaf scald as a disease can cause major disruption to the production of disease-free seed plans. In some cases, plots may have to be relocated to areas isolated from known sources of disease.

Symptoms

The plants infected with the leaf scald pathogen do not always display external symptoms. These plants are referred to as being latently infected, and the mechanism of latent infection is not understood so far. There are also cases of apparent recovery in which symptoms subside and do not become visible until ratoon crop regrowth or after planting infected seed cane. However, during this apparent recovery, the disease is in a period of latent infection in the affected cane.

Leaf scald is further complicated by the fact that it may be manifested in a chronic phase (a progressive increase of disease severity) or an acute phase (sudden appearance of the disease and plant death). The chronic phase is characterized by several external symptoms. The most typical symptom is a white pencil line streak about 1–2 mm wide on the leaf that extends from the midrib to the leaf margin running parallel to the veins. A diffuse yellow border of varying widths runs parallel to the pencil line streak. The pencil line may have areas of reddish discolouration along part of its length. As the disease progresses, necrosis develops from the leaf tip or leaf margin, and finally extends the entire leaf. Leaves look burned and curl inward, giving the foliage a scalded appearance, hence the name for the disease. Ultimately, the entire plant dies. Sprouting of lateral buds of the matured canes occurs in an acropetal fashion.

The disease can also cause shoots to be stunted and wilted. Usually, affected leaves turn a dull blue-green colour before dense browning (a late symptom of the disease). On mature stalks, the spindle leaves become necrotic from the tips and moderate to profuse side shoots to develop. Side shoots first appear at the bottom of the stalk and progress upward. These side shoots usually show scalding and/or white pencil lines. The side shoots often die while quite small (<18 inches or 46 cm).

Causal organism: *Xanthomonas albilineans*

In the early stages of infection, the leaf scald bacterium is restricted to the xylem elements of the vascular bundles in the white pencil line streaks. It is generally not found in the surrounding chlorotic leaf tissues. A phytotoxin called albicidin has been isolated from chlorosis-inducing strains of *X. albilineans*. This phytotoxin inhibits chloroplast differentiation and thus disrupts photosynthesis. In the late stages of infection, the pathogen exits the xylem and invades other tissues, causing the appearance of lysigenous cavities in the stalk (Mensi *et al.*, 2014). The variants of the pathogen have been identified worldwide and there are at least three serological strains of the bacterium, and several variants in virulence have also been reported. The yield of stalks that are dead or have necrotic tops and leaves with numerous side shoots is decreased to 20%–30% of that of symptomless stalks.

Disease cycle

Xanthomonas albineans can spread from inoculum sources to contaminated fields of sugarcane and affect healthy sugarcane under the impact of various climatic conditions (Champoiseau *et al.*, 2009). The pathogen then colonizes the surface of the leaf, enters through stomata, and progresses within the xylem, and symptoms may appear in infected plants. Pathogen can move into the stalk and then infect the stool showing scalding on the leaves (Daugrois *et al.*, 2011). The pathogen can be transmitted mechanically by harvesting equipment and infected cane setts.

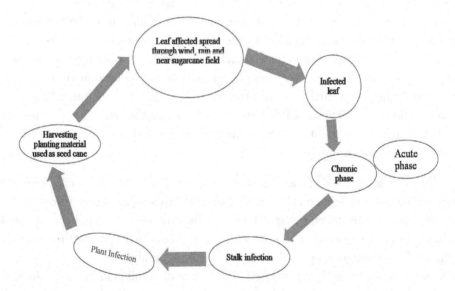

Epidemiology

The amount of damage caused by leaf scald appears to be influenced by environmental conditions. Periods of stress such as drought, waterlogging, and low temperature is reported to increase disease severity. Leaf scald is a systemic disease that may be inconspicuous (latent) for lengthy periods of time, and infected seed cane is a major cause of disease spread (Ricaud and Ryan, 1989). The equipment used during operations like cutting knives and machinery is an important source of infection.

The pathogen survives in stubble but does not appear to survive for long periods of time in soil or un-decomposed cane trash.

The secondary source of inoculums for the disease includes alternate hosts such as elephant grass (*Pennisetum purpureum*) which offer another means of pathogen survival. The aerial spread of leaf scald has also been reported in Florida which was caused by a new strain of pathogen having better survival on leaves (Davis *et al.*, 1997).

Integrated disease management

a. Control of leaf scald involves a combination of deploying resistant varieties and replacement of susceptible varieties with disease-free or approved seed schemes and hygiene. The disease shows a period of latency, therefore the resistant varieties have been to be regularly monitored for infection.

b. The seed cane (mother or nucleus seed) can be freed from the disease by long soaking in cold water (40 hours) followed by hot water treatment at 50°C for 3 hours. If leaf scald is common in an area, the mother plot and approved seed plot can be relocated to a low-risk area.

c. The mechanical spread of the pathogen can be prevented by disinfection/ sterilization of all cane cutting knives, including those on mechanical harvesters. Disinfection of the knives can be accomplished by cleaning and immersing for several minutes in a suitable antiseptic solution, such as Lysol, alcohol, or a diluted solution of bleach. There are no known chemical or biological controls for this disease.

10. Ratoon stunting disease

Economic importance

Ratoon stunting disease (RSD) was first detected in 1944 in the ratoon crop of Q28, in Queensland (Australia) by Steindl who coined the terminology ratoon stunting for the disease. The disease has been reported in Australia, USA, India, Brazil, Florida, China, Fiji, Philippines and South Africa (Antoine, 1958; Spaull and Bailey, 1993; Mayeux *et al.*, 1998; Rao and Singh, 2000; Vishwnathan, 2001; Dela *et al.*, 2002; Xu *et al.*, 2008; Johnson and Tyagi, 2010).

Symptoms

Diseased clumps usually display stunted growth, reduced tillering, thin stalks with shortened internodes and yellowish foliage. The infected canes show orange-red vascular bundles in shades of yellow at the nodes. The major loss is enhanced by stress, particularly moisture stress and reduced stalk height and diameter (Gao *et al.*, 2008; Comstock and Gilbert, 2009). The RSD pathogen seems to impair normal metabolic growth processes in the roots of infected canes. Poor germination of RSD-infected buds has been attributed to the absence of acid invertase (Madan *et al.*, 1986). The inhibition of acid invertase by the RSD pathogen could be probably due to a factor which either inhibited acid invertase directly or interfered with the synthesis of DNA-dependent m-RNA synthesis. The factor is probably thermolabile and the inhibition is reversible since the MHAT of the infected sets circumvents the inhibitory effects of the causal agent on enzyme activity.

Causal organism

Clavibacter xyli sub sp. *xyli*

The pathogen (*Clavibacter xyli* sub sp. *xyli)* is an RLO known to be present in the xylem cells of infected plants. They are small, thin, rod-shaped or coryneform (0.15 to 0.32 μm wide and 1.0-2.7 μm long) and Gram-positive.

Disease cycle

The primary spread is through the use of diseased setts. The disease also spreads through harvesting implements contaminated with the juice of the diseased canes. Maize, sorghum, sudan grass and cynodon serve as collateral hosts for the pathogen.

Epidemiology

The disease flares up under drought conditions and the impact of the disease is more in the ratoon crop as compared to the plant crop (Agnihotri, 1990). The pathogen can be spread during propagation in infected cuttings from field to field or mechanically on equipment from plant to plant within a field. The titer of bacteria in xylem sap increases during the growing season, so the most important periods of spread occur during planting and harvest.

Integrated disease management

a.   Almost all the sugarcane varieties are susceptible to RSD to some extent. The diseased setts or contaminated cutting implements such as cane knives, harvesters or planting machines can spread RSD. Therefore, selections of setts from disease-free fields or from disease-free commercial nurseries along with disinfection of contaminated cutting implements are very important.

b. Removal and burning of the infected clumps is also a suggestive measure to prevent the spread of disease. The practice of volunteer free fallow period is also an opportunity to break the pest and disease cycle and eliminate the disease from the block.

c. Heat treatment of the setts at 50-52°C for one-two hours before planting will inactivate the pathogen, as specified for grassy shoot disease.

11. Red stripe

Economic importance

Red stripe or top rot earlier was a minor bacterial disease of sugarcane and is now an important disease. The disease was first observed in Hawaii by Lyon (1922). After that disease was reported in many cane-growing areas (Ricaud et al., 1989). Tryon (1923) reported the top rot condition of sugarcane in Queensland and later reported it in 50 sugarcane growing countries. Disease incidence may range from 8-80% in China in different varieties under different climatic conditions (Shan et al., 2017). In India, the disease was first reported by Desai in 1933 and was subsequently reported in several other states on many sugarcane varieties (Agnihotri, 1990). Chona and Rao (1963) recorded the incidence of Red stripe disease in an epidemic form around Delhi and the eastern districts of Punjab. Chaudhary et al., (1999) reported a 38 and 40% reduction in cane yield and recoverable sugar respectively by the disease. In Rio de Janerio (Brazil), field losses of up to 15% or more have been reported (Vesminsh et al., 1978). Kumar et al., (2014) recorded the incidence of top rot disease (54.3-56.0% and 54.3%) in Amritsar (Punjab, India) district on a variety of CoJ 85 in different years.

Symptoms

Red stripe showed symptoms on the basal part of the young leaves of the spindle as water-soaked lesions and long narrow chlorotic streaks near the midrib. Mostly the lower half of the leaf is more affected than the upper half. These stripes maybe 0.5 to 1 mm in width and 5 to 100 mm in length (Rangaswami and Rajagopalan, 1973). In many countries like Hawaii (Martin, 1938), Java (Bolle, 1929), and Taiwan (Okabe, 1933) young ratoons were more affected than plant canes. Fors (1978) observed red-coloured blotches on stalks in the region of root primordia extending towards the internodes in the form of thick red lines. Red stripe occurs mostly on the young and middle-aged leaves, rather than on the older leaves of the plant. The disease may attack the youngest leaves which are partially unrolled and, if incidence is severe, causes a top rot. Canes infected with top rot emit a characteristic unpleasant odour (Martin and Wismer, 1961). In North India symptoms were reported to appear from July to August and the infection may result from stem, bud or leaf (Rana and Shukla, 1968).

Causal organism

The disease is caused by *Acidovorax avenae* subsp. *avenae* bacterium. The pathogen is a Gram-negative bacterium, rod-shaped, occasionally forms chains, measures 0.7 x 1.6 µm, and does not produce endospores. Earlier, Cottrell-Dormer (1932) showed that top rot and red stripe were a manifestation of the same disease and were caused by the same organism (*Pseudomonas rubrilineans*). Orian (1956) reported disease with symptoms similar to those of red stripe (*Pseudomonas rubrilineans*) on Mauritius in 1954 on Barbados varieties B3337 and B37161. Patro *et al.,* (2006) reported that the pathogen associated with the top rot phase of sugarcane red stripe disease in India is *Acinetobacter baumannii*.

Disease cycle

The presence of sugarcane throughout the year helps the survival of the pathogen which effects the young tissues of the host. The pathogen also infects and survives in jowar (*Sorghum vulgare*), bajra (*Pennisetum typhoides*), maize (*Zea mays*), etc. Primary infection can be caused in two ways. First, the setts taken from diseased plants may carry bacteria with them and the plants developed from such setts may show disease development in them. Second, the bacteria surviving on sugarcane of neighbouring fields and hosts, ooze out the slimy substances which, when disseminated by wind and rain splashes, cause primary infection on healthy sugarcane plants. Bacterial pathogens produced as a result of primary infection usually spread in the field by wind and rain-splashes and cause secondary infections during the growing season.

Epidemiology

The bacterium can infect all parts of the plant through stomatal openings and wounds. The primary and secondary transmission takes place by seed cuttings and rain, water and insects respectively. The secondary transmission takes place by wind and rain (Martin and Wismer 1961).

Integrated management

Patro *et al.*, (2006) reported that the causal organism of the red stripe was highly resistant to Ampicillin, but was susceptible to the rest of the antibiotics tested (Streptomycin, Kanamycin and Rifampicin). Dange and Payak (1974) proved that *Pseudomonas rubrilineans* is quite sensitive to Terramcin as well as Streptocycline. However, Hussnain *et al.*, (2011) reported that Ampicillin and Vancomycin at 75 and 25µg/ml, respectively were effective against *Acidovorax avenae* subsp. *avenae* when compared with other saprophytic bacteria.

12. Grassy shoot disease

Economic importance

Sugarcane grassy shoot (SCGS) is one of the most important diseases of sugarcane in India after fungal diseases (Rao and Dhumal, 2002). In India, SCGS disease has been reported in Punjab, Uttar Pradesh, Haryana, Bihar, West Bengal, Madhya Pradesh, Andhra Pradesh, Karnataka and Tamil Nadu (Vasudeva, 1955). The disease was first observed by Barber (1919) and reported by Chona *et al.* (1958) from Belapur (Maharashtra). The grassy shoot disease has been reported to contribute to losses of 5 to 20 per cent in the main crop and these losses are up to 100% in the ratoon crop (Rao *et al.*, 2008, Marcone *et al.*, 2004, Vishwanathan & Rao, 2011). Primarily SCGS infected plants are limited in number, but the incidence increases up to 60-80 % in ratoon crops through secondary spread by insect vectors (Srivastava *et al.*, 2006). A higher incidence of SCGS has been recorded in some parts of Southeast Asia and India, resulting in a 100% loss in cane yield and sugar production.

Symptoms

The disease appears nearly two months after planting and is characterized by the production of numerous lanky, adventitious tillers from the base of the affected shoots giving the plant a bushy appearance bearing pale yellow or chlorotic leaves which remain thin, narrow, and reduced in size (Chona *et al.*, 1958, Sarosh *et al.*, 1986, Rishi & Chen 1989). The plants appear bushy and 'grass-like' due to a reduction in the length of internodes, and premature and continuous tillering. There is the formation of white leaves by leaf chlorosis and proliferation of tillers, excessive tillering and stunting of the plants gives the plant a grassy appearance (Nasare *et al.*, 2007) and hence the name grassy shoots disease. The affected clumps are stunted by the premature proliferation of auxillary buds. The cane formation rarely occurs in the affected clumps, if formed, thin with shorter internodes having aerial roots at the lower nodes and is not millable. The buds on such canes are usually papery and abnormally elongated. If the attack is light, one or two weak canes may be formed. Most of the stools die after the monsoon. The severely diseased clumps remain stunted and may produce one or two weak canes. The disease is particularly pronounced in the ratoon crop giving the appearance of a field full of perennial grass.

Pathogen

The disease is caused by phytoplasma "*Candidatus* Phytoplasma sacchari". The pathogen produces two types of bodies which are seen as ultrathin sections of phloem cells of infected plants. The spherical bodies of 300-400 nm diameter and filamentous bodies of 30-53 mm diameter in size. The phytoplasma usually colonizes sieve tube elements in phloem systemically, thereby blocking the movement of photosynthates from leaf to stalk and root. This results in poor growth of sugarcane and reduced sugar recovery. The altered metabolism in the host induced by the pathogen infection causes different types of symptoms such as albino, excess tillering, leathery leaves, bud sprouting etc.

Disease cycle and transmission

The vector(s) responsible for the natural spread of SCGS has not been identified. According to some reports, the disease is primarily spread by infected seed setts while secondary infection may involve insect vectors especially leaf hoppers, plant hoppers and psyllids from the family Cicadellidae, Fulgoroidea and Psylloidea in a persistent propagative manner (Vasudeva, 1960, Singh, 1969, McCoy *et al.*, 1989, Srivastava *et al.*, 2006). Also, there are reports on transmission by three different species of aphids (currently named *Rhopalosiphum maidis* (Fitch), *Melanophis sacchari* (Zehntner) and *Melanophis idiosachhari* as well as by *Proutista moesta* (Westwood), a fulgorid (Chona *et al.*, 1960, Edison *et al.*, 1976). The leafhopper has been reported to transmit SCGS phytoplasma in India. In India, sugarcane grassy shoot disease has been reported to be transmitted by leafhopper (Edison, 1973, Rishi & Chen, 1989, Tran Nguyen *et al.*, 2000, Singh *et al.*, 2002, Srivastava *et al.*, 2006). Singh *et al.*, (2002) and Srivastava *et al.*, (2006) reported that nymphs of leaf hopper *Deltocephalus vulgaris* were more efficient than adults in transmitting the SCGS phytoplasma. Mechanical transmission through cutting knives etc. is doubtful though transmission through dodder plants (*Cuscuta campestris*) has also been reported. The disease increases in successive ratoon crops. Sorghum and maize serve as natural collateral hosts.

Integrated disease management

In SCGS disease, the primary concern is to prevent the disease rather than treat it.

a.   Healthy planting material

As vegetative propagation in sugarcane favours harbouring of grassy shoot phytoplasma, adequate care should be taken while selecting seed canes. The crop with a 2% GSD incidence is unsuitable for seed purposes. Therefore, the seed cane plots are to be monitored periodically for the disease at regular intervals and the seed canes should be selected from such carefully monitored fields. The tissue culture seedlings derived through meristem ensure freedom from the pathogen. However, it is advised to select a mother plant (for meristem culture) from well-maintained fields. The three-tier seed production system is a viable procedure to maintain the area disease-free and obtain healthy seeds.

2.   Heat therapy

At times when disease-free materials are not available, heat therapy (either aerated steam or moist-hot air) needs to be employed to eliminate the phytoplasma pathogen. A temperature regime of 50- 52°C for one hour is ideal to inactivate the pathogen. If the varieties are sensitive at 52°C, treatment time may be increased to 50°C for 2-3 hrs to inactivate the pathogen. During the course of heat treatment and further handling of the setts, are should be taken to minimize the loss to the buds. Always the heat-treated setts should be treated in a fungicide solution to protect the setts from soil-borne pathogens.

13. Yellow leaf disease

Economic importance

Yellow leaf disease (YLD) of sugarcane affects production significantly in all sugarcane-growing areas of the world. Yellow leaf disease (YLD) of sugarcane was first reported in Hamakua (Hawaii) on variety H65- 0782 in 1989 as yellow leaf syndrome (Schenk, 1990) and subsequently from the United States mainland (Comstock *et al.*, 1994) and many other sugarcane growing countries. The disease is reported worldwide in more than 30 countries (Lockhart 1996). It was first reported in India in 1999 and in recent years it has attained epidemic proportions, seriously affecting sugarcane production in the country. In India, the disease is prevalent in major sugarcane-growing states like Andhra Pradesh, Karnataka, Tamil Nadu and Madhya Pradesh (Vishwanathan and Rao, 2011). The incidence of SCYLV in commercial fields can reach 100% in susceptible cultivars, and the disease can cause significant yield losses in susceptible cultivars even if infected plants do not exhibit the disease symptoms (Viswanathan 2002, 2016).

Symptoms

The disease primarily affects 5-6 months of crop resulting in complete yellowing of leaf midrib, yellow laminar discolouration on both sides of the midrib and in severe cases necrosis of the leaf lamina from the leaf tip spreading downwards along the midrib. There are reports of reddish discolouration as well in some symptomatic plants along with drying of spindle and leaves. In highly susceptible varieties, affected plants show a bushy appearance of leaves at the crown due to internode shortening during the maturity stage under Indian conditions. Unrestricted movement of virus-infected seed canes and repeated use of infected seed cane resulted in a severe outbreak of the disease in different states in the country. Such fields show stunted growth accompanied by extensive foliage drying.

Causal agent

Sugarcane Yellow Leaf Virus

SCYLV is a Polerovirus (Family Luteoviridae) evolved by recombination between the ancestors of Luteovirus, Polerovirus and Enamovirus (Moonan *et al.*, 2000 and Smith *et al.*, 2000). The SCYLV is known to be transmitted by aphids, *Melanaphis sacchari* and *Rhopalosiphum maidis*, in a semi-persistent manner. The SCYLV is a positive-sense, ssRNA virus of about ~6 kb in size (Moonan *et al.*, 2000). SCYLV genome is composed of six open reading frames (ORFs) viz., ORF0, 1, 2, 3, 4, and ORF5 with the three 5'-untranslated regions (Moonan and Mirkov, 2002; Smith *et al.*, 2000). The SCYLV is known to have ten genotypes documented based on whole genome characterization and designated as BRA (Brazil), CHN1, CHN2 and CHN3 (China), CUB (Cuba), HAW (Hawaii), IND (India), PER (Peru), COL (Colombia), and REU (Reunion Island).

There is strong evidence that Sugarcane yellow leaf phytoplasmas (SCYP) are also associated with leaf yellowing in some countries *viz;* South Africa, Mauritius and in Cuba. Several workers reported the association of phytoplasmas with YLD (Cronje *et al.*, 1998, Cronje and Bailey, 1999, Marcone, 2002, Parmessur *et al.*, 2002; Arocha *et al.*, 2005) from nine African countries. After detailed studies on the aetiology of YLD in Mauritius, Aljanabi *et al.* (2001) concluded that either SCYLV or SCYP or their combination is associated with YLD symptoms. In India, the first report of the 16SrXII group of phytoplasma in the YLD-affected sugarcane genotypes was made by Gaur *et al.*, 2008; Rao *et al.*, 2012 and 'Ca. phytoplasma asteris'(subgroup: 16SrI-B) was detected in the two YLD-affected sugarcane varieties i.e., CoLk94184 and CoSe 92423 by Kumar *et al.* (2015).

Virus vector interaction

The yellow leaf disease primarily spreads through insect vectors like *Melanaphis sacchari* and *Rhopalosiphum maidis, R. rufiabdominalis* in case of SCYLV (Edon-Jock *et al.*, 2007; Schenck and Lehrer, 2000). The virus is phloem–limited and resides in phloem parenchymatous tissues of plants and spread by insect vectors (aphids) in a persistent, circulative, and nonpropagative means and cannot be transmitted by artificial sap inoculation (Rochow, 1982). The transmissibility of SCYLV is preserved without replicating inside for the complete life of aphids and is not even lost during their moulting. The viral particle circulates into the insect gut, hemolymph and salivary glands for further transmission (Bragard *et al.*, 2013; Pinheiro *et al.*, 2015). Members belonging to the family Luteoviridae follow the transcytotic dissemination pathway (Gutierrez *et al.*, 2013). After the entry of virions into the aphid mouthparts, are moved through the foregut, midgut and hindgut. Initially, virion interactions occur with the epithelial cells of the gut having receptors which facilitate the adherence to midgut and or hindgut via endocytosis. Afterwards, with the help of the exocytosis phenomenon virus particles travel to the haemocoel, which passes through, subsequently, virions reach to salivary glands for its transmissibility *via* saliva while probing for the plant sap (Garret *et al.*, 1996; Gildow, 1993; Gray *et al.*, 2014).

Integrated disease management

a. The management of SCYLV is difficult due to its vector-borne nature and transmission through infected seed cane. The severity of the disease is also known to vary in different varieties cultivated in different agro-climatic conditions. However, the disease can be effectively managed only through the integration of cultural, chemical, and biological methods, host plant resistance, pathogen-derived resistance, RNAi silencing and appropriate diagnostic techniques that need to be adopted.

b. Among the conventional management strategies, healthy seed production by a three-tier system must be emphasized (Singh and Singh, 2015). Adoption of wide row spacing and early planting can help in the alleviation of YLD losses (Palaniswami *et al.*, 2014; Viswanathan *et al.*, 2017). The practice of multiple vegetative propagations of a single seed should not be practiced.

c. Disease surveillance through remote sensing techniques for monitoring and identification of YLD-affected sugarcane fields is also highly recommended.

d. Biological control for aphid control through entomopathogenic agents like *Verticillium lecanii* (Hall, 1987) and predators including so*lanum nigrum* (Mulsant), *Allograpta exotica* (Wiedemann), *Coleomegilla maculate fuscilabris* (Mulsant), *Hippodamia convergens* (Guerin), *Diomus terminates* (Say), *Lysiphle bustestaceipes* (Cresson), *Micromus subanticus* (Walker), *Chrysoperla externa* (Hagan), and *Cycloneda sanguinea* (L.) (Hall, 1987, 1988; White *et al.*, 2001). The sugarcane aphid vectors can also be managed through the use of yellow sticky traps (Satyagopal *et al.*, 2014).

e. The aphid *M. sacchari* population can be effectively reduced up to 83.3%, 80.2%, 79.7%, and 77.5%, respectively by application of 0.05% solution of dimethoate 30 EC, 0.07% endosulfan 35 EC, 0.04% monocrotophos 36 WSC or 0.05% chlorpyriphos 20 EC (Balikai, 2004; Viswanathan *et al.*, 2017a). However, the application of insecticide sprays to manage aphids is not feasible when the crop in the field is more than five to six months old, for which automatic aerial sprays are helpful.

f. Apical meristem tip culture, auxiliary bud culture and leaf roll callus culture techniques have been found effective in the management of YLD in commercial and noble sugarcane cultivars (Chatenet *et al.*, 2001; Fitch *et al.*, 2001; Parmessur *et al.*, 2002). Meristem tip culture and auxiliary bud culture are the most advantageous and common methods for virus elimination from all susceptible varieties, due to the advantage of the fact that, meristematic tissue remains free from the virus (Fitch *et al.*, 2001; Parmessur *et al.*, 2002). Viswanathan *et al.* (2018) efficiently validated the impact of virus-free planting materials of sugarcane variety, Co 86032 through the meristem tip culture technique. Thus, the use of improved diagnostic techniques and production of virus-free plants through meristem tip culture have been found as a feasible strategy to manage YLD and it facilitates the continuous supply of healthy seedlings to the sugarcane growers for enhanced and sustained sugarcane productivity in India. Production of disease-free sugarcane seedlings is the prerequisite for enhanced production and productivity.

Plant-Parasitic Nematodes of Sugarcane

1. Lesion Nematodes (*Pratylenchus* species)

Economic importance

The most important PPNs of sugarcane crops have cosmopolitan occurrence and are of the migratory endo-parasitic type of feeding. Various species of *Pratylenchus* have been recorded from the sugarcane fields *viz.*, *brachyurus*, *coffeae*, *goodeyi*, *pratensis* and *zeae*. Among them, *coffeae*, and zeae have the most widespread distribution, observed in high population and causes severe economic damage to sugarcane cultivation. All the stages (four juveniles, male and female) of the nematode are infective.

Symptoms

Above-ground symptoms of nematode infection are non-specific and are of general wilting. Infected cane shows dwarfing, and stunting symptoms which gradually leads to decay of primary roots, wilting and death. The most characteristic symptoms of lesion nematodes are dark black-brown coloured lesions on roots. The nematode enters the root through cortical parenchymatous tissues by rupturing the root and causes damage to the neighbouring cells by puncturing the stylet. Being migratory endoparasite, nematodes with the help of oesophageal secretions and stylet puncturing, travel inside the root and cause necrotic brown-black spots (lesions) on roots. At first, lesions are small, elongated, water-soaked spots that finally coalesce and gradually girdle the roots. Besides root damage caused by nematode alone, the crop is prone to susceptible to soil-borne pathogens (*Fusarium* sp., *Pythium* spp., *Cephalosporium* sp.)as a secondary infection which further aggravates the damage and reduction in the yields (Rashid, 2001).

2. Lance Nematode (*Hoplolaimus* species)

Economic importance

Most commonly occurred PPNs in sugarcane cultivation next to *Pratylenchus* species. Till now, lance nematode has assumed as a minor PPN problem in agriculture crops, but its significant presence as a versatile feeding style (migratory to endo-semi endo and ectoparasites) in poly-specific nematode communities and occurred in high population, the status of minor pest modifies to economically significant PPN in sugarcane crop. The important species reported *viz.*, *galeatus*, *seinhorstii*, *pararobustus* and most dominant *indicus*. *H. indicus* is most widespread in tropical and sub-tropical Indian sugar belts, particularly in the command areas of sugar factories. *H. indicus* was first described from soil samples nearer to sugarcane roots from Punjab, India.

Symptoms

General crop stunting and loss of weight in fresh and dry tops, foliage dropping etc. are the general sign of nematode infestations. The roots show prominent discolouration, devoid of laterals, and sparse and stubby appearance which gradually succumbs to cell necrosis. The cortical and parenchymatous root tissues are badly attacked. At heavy infestations, cane stool/clump can easily be pulled out.

Life cycle

Hoplolaimus spp. are of having versatile feeding habits (from root browsers to migratory Ecto/endoparasite). All the stages (four juveniles, adult male and female) are infective. Reproduce by amphimixis. The migration of nematodes in the soil is oriented towards the stimuli exudated from roots and especially towards the root tip region for entry. The nematode explores the root region for entry, penetrates the stylet, punctures the cortical cells, and retrieves the food material from the host crop. Histologically, the infected root cells become thickened, devoid of cytoplasm, formation of feeding cavities, and membrane modification at the ruptured site caused by nematode infection. Being migratory endoparasite and ability to cause extensive root damage, the lance nematode infection allows/provide entry for other phytopathogens *viz.*, *Fusarium* sp., *Pythium* spp. etc. The combined damage caused by nematode and soil-borne pathogens is observed to be more extensive than individual effects.

3. Root-knot nematode (*Meloidogyne javanica*)

Economic importance

Another genus of importance to sugarcane is *Meloidogyne* spp. In India, *M. javanica* occurs in the dominant areas of sugarcane. Although RKN is considered the most economically important nematode in the world, it is not that significant in India on sugarcane and is found only in specific pocket areas of sugarcane in south India, western Uttar Pradesh and a few locations in Kerala.

Symptoms

In highly cultivated areas of sugarcane, nematode-infested roots show a few tiny-sized pin-shaped galls concomitantly associated with jelly-like egg mass around them. The infected root gradually decays due to the combined action of nematode and root rotting fungi. The foliar symptoms are very much general to stunting, yellowing and discolouration of leaves. Heavily infected sugarcane stool can be easily pulled out at the young stage of the crop.

Life cycle

The juvenile enters the root through the region of root proliferation. Heavy nematode infection can cause cessation of root growth by affecting meristematic root tissue. Giant cell formation, maturing females and expanding the adjacent root tissue exert pressure on translocation of water and nutrients thereby affecting malformation or distortion in uptake. Distinct root curvature is a distinct feature observed in RKN infection.

4. Spiral Nematode (*Helicotylenchus* spp.)

Economic importance

This nematode has been found in high numbers in random sampling from sugarcane crops. More than 21 species of this genus have been reported on sugarcane all over the world. The most dominant is *dihystera* and *multicintus*. Among these, *H. dihystera* was reported first time on sugarcane in 1893 from Hawaii.

Symptoms

The most characteristic aerial symptoms are highly chlorotic leaves upon nematode infestation. At a higher population of nematodes on sugarcane roots, due to feeding externally on the root surface, the epidermis is badly damaged and sloughed off. The root system exhibits blunt, deformed and stunted in size, gradually decay and is finally prone to other soil-borne pathogens.

Stunt-stylet Nematode (*Tylenchorhynchus* species)

Most widely occurred with sugarcane rhizosphere. Important species are *brevilineatus*, *martini*, *elegans*, and *crassicaudatus*. The ectoparasitic style of feeding is generally observed with this nematode. Primarily, feeding on the root surface, root hairs and epidermal cells. A higher population of nematodes occurred frequently but no prominent symptoms on roots. However, general stunting of sugarcane is seen in the higher population of nematodes.

In addition to this, other mild nematode ectoparasites of sugarcane were also reported. *Rotylenchus renifromis* was also found in a significant number however no adults (females) have been reported from roots. Probably, nematode completes their life cycle on weeds present in sugarcane fields. Dagger nematode, *Xiphinema*spp also found in low population

Nematode Management:

1. Crop rotation with dicot plants will help to reduce the PPN population by a principal to cutting off the continuous food supply and staving the nematode stages.

2. Application o press mud cake (PMC) at a rate of 1 ton/ha influence the nematode community and drastically reduce the PPN population by augmenting the biocontrol fauna in soil.

3. Weed management is very much crucial for nematode management as most of the PPN harbours their stages on weed host and continue their life cycle to cause infestation on sugarcane.

4. Improved agronomic practices (strip cropping, intercropping) will help to build up nematode antagonists that will ultimately influence on PPN population.

Future Thrusts

The sugarcane diseases are the major hindrance in realizing the maximum crop production across the growing areas of several countries. Understandings of host-pathogen interaction at the molecular level are needed to elucidate the evolution and diversity of pathogenicity mechanisms which may help in the formulation of novel disease management strategies against pathogens. Management of disease through an integrated disease management strategy is the only viable option in smut disease control. The biological control of this disease is an option to reduce the dependence on chemical fungicides. The endophytic bacterial community associated with sugarcane also has the potential to manage the disease. Identification of regional pathotypes across the country and their use in the development of smut resistant varieties for the particular region is the need of the hour. Host plant resistance combined with clean cultivation practices would lead to the successful management of smut in sugarcane. With the available molecular tools, conventional breeding can strengthen through the identification of suitable markers to the gene of interest. The quick and precise detection of the pathogens in seed cane is an emergent need to restrict the pathogen spread from one region to another. The rapid diagnosis followed by strict quarantine regulations would limit the disease in a region. The limited information on the epidemiology of most of the diseases paves the way for new areas of research. The study will help in the understanding of the interaction between change in climate and disease spread which in turn strengthens the decision-support systems through disease forecast. The novel strategies like induction of systemic resistance against pathogens should be thought of efficient management. Further, the use of artificial intelligence for diagnosis, prediction and identification of candidate defense genes against pathogens can be looked upon in the future.

References

Chatenet, M., Delage, C., Ripolles, M., Irey, M., Lockhart, B. E. L. & Rott, P. (2001). Detection of Sugarcane yellow leaf virus in quarantine and production of virus-free sugarcane by apical meristem culture. *Plant Dis.* 85:1177-1180.

Chaudhary, M. A., Ilyas, M. B., & Khan, M. A.(1999). Effect of top rot of sugarcane on yield quality, relationship of temperature and humidity with disease development and screening for varietal resistance *Pakistan J. Phytopathol,* 11: 126-129.

Chona, B. L., & Nariani, T. K. (1954). Investigations on the survival of the sugarcane red rot fungus in the compost. *Indian Phytopath* 5: 151-157.

Chona, B. L., & Rao, Y. P. (1963). Association of *Psedomonas rubrilineans* (Lee *et al.,*) Stapp. with the red stripe disease of sugarcane in India. *Indian Phytopath,*16:392-93.

Chona, B.L., & Srivastava, D.N. (1952). The perithecial stage of *Colletotrichum falcatum* Went in India. *Indian Phytopath* 5:158-160.

Chona, B.L., Capoor, S.P., Varma, P.M., & Seth, M.L. (1958). Grassy shoot disease of sugarcane, *Indian Phytopath.* 13: 37-47.

Chona, B.L., Capoor, S.P., Verma, P.M., & Seth, M.L. (1960). Grassy shoot disease of sugarcane, *Indian Phytopath.* 30: 37-47.

Comstock, J.C. and Gilbert, R.A. (2009). Sugarcane Rat Stunting Disease, SS-AGR-202: 1-3.

Comstock, J.C., Irvine, J.E. & Miller, J.D. (1994). Yellow leaf syndrome appears on the United States mainland. *Sugar J.* 56: 33–35.

Costa, M. M., Melo, M. P., Guimaraes, E. A., Veiga, C. M. O., Carmo Sandin, F., Moreira, G. M., et al & Pfenning, L. H. (2019). Identification and pathogenicity of Fusarium species associated with pokkah boeng of sugarcane in Brazil. *Plant Pathol,* 68(7), 1350-1360.

Costa, M.M., Melo, M.P., Guimarães, E.A., Veiga, C.M.O., Carmo Sandin, F., Moreira, G.M., & Pfenning, L.H. (2019). Identification and pathogenicity of Fusarium species associated with pokkah boeng of sugarcane in Brazil. *Plant Pathol*, 68(7): 1350-1360.

Cottrell-Dormer, W. (1932). Red-Stripe disease of sugarcane in Queensland. *Bull Bur Sugar Exp Stn Div, Pathology* 3: 25- 59.

Cronje, C.P.R & Bailey, R.A (1999). Association of phytoplasmas with yellow leaf syndrome of sugarcane. In: Proceedings of the 23rd Congress of the International Society of Sugarcane Technologists, New Delhi, India, 1999, 373-381.

Cronje, C.P.R., Tymon, A.M., Jones, P. & Bailey, R.A. (1998). Association of a phytoplasma with a yellow leaf syndrome of sugarcane in Africa. *Ann. Appl. Biol.* 133:177-186.

Dange, S. R., Payak, M. M. (1974). Sensitivity of *Pseudomonas rubrilineans* to antibiotics and fungicides. *Hindustan Antibiot Bull*, 16(2-3):89-92.

Daugrois, J.H., Dumont, V., Champoiseau, P., Costet, L., Boisne-Noc, R. & Rott, P. (2003). Aerial contamination of sugarcane in Guadeloupe by two strains of Xanthomonas albilineans. *Eur. J. Plant Pathol.* 109: 445–458.

Davis, M.J., Rott, P., Warmuth, C.J., Chatenet, M. & Baudin, P. (1997). Intraspecific genomic variation within Xanthomonas albilineans, the sugarcane leaf scald pathogen. *Phytopathol*, 87: 316–324.

DelaCueva, F.M., Natural, M.P., Bayot, R.G., Mendoza, E.M.T & Ilag L.L. (2002). Geographic distribution of ratoon stunting disease of sugarcane in the Philippines. *J Tropic Plant Pathol.* 38: 11-18.

Duttamajumder, S. K. (2008) Red Rot of Sugarcane. Indian Institute of Sugarcane Research, Lucknow, India.

Edison, S. (1973). Investigations on the grassy shoot disease of sugarcane, Ph.D. thesis Tamil Nadu Agric. Univ., Coimbatore.

Edison, S.K., Ramakrishnan & Narayanasamy, P. (1976). Comparison of grassy shoot in India with white leaf disease of Taiwan in sugarcane. *Sugarcane Pathol. Newsl.* 17: 30-33.

Edon Jock, C., Rott, P., Vaillant, J., Fernandez, E., Girard, J.C. & Daugrois, J.H.

(2007). Status of Sugarcane yellow leaf virus in commercial fields and risk assessment in Guadeloupe. *Proc. Int. Soc. Sugar Cane Technol.* 26:995-1004.

Fitch, M.M.M., Lehrer, A.T., Komor, E. & Moore, P.H. (2001). Elimination of Sugarcane yellow leaf virus from infected sugarcane plants by meristem tip culture visualized by tissue blot immunoassay. *Plant Pathol.* 50:676-680.

Fors, A.L. (1978). Red stripe in Central America. *Sugarcane Pathology Newsletter,* 21: 25-26.

Ganguly, A., & Jha, T. N. (1964). Improved inoculation method for testing sugarcane varieties for resistance against wilt disease. *Sci Cult,* 30: 456-458.

Gao, S.J., Pan, Y.B., Chen, R.K., Chen, P.H., Zhang, H. & Xu, L.P. (2008). Quick detection of Leifsonia xyli subsp. xyli by PCR and nucleotide sequence analysis of PCR amplicons from Chinese Leifsonia xyli subsp. xyli isolates. *Sugar Tech.* 10: 334-340.

Garret, A., Kerlan, C. & Thomas, D. (1996). Ultrastructural study of acquisition and retention of Potato leafroll luteovirus in the alimentary canal of its aphid vector, *Myzus persicae* Sulz. *Arch. Virol.* 141:1279-1292.

Geh, S.L. (1973). Current status of diseases and pests of sugarcane in West Malaysia (pp. 4 - 6). Mardi Report

Giatgong, P. (1980). Host index of plant diseases in Thailand. (2ndedn), Dept. of Agriculture, Ministry of Agriculture and Cooperatives, Bangkok, Thailand.

Gildow, F.E. (1993). Evidence for receptor-mediated endocytosis regulating luteovirus acquisition by aphids. *Phytopathol* 83:270-277.

Goswami, S.K., Singh, D., Joshi, D., & Singh, S.P. (2020). An insight into sugarcane wilts in India. *Agric Res J,* 57(5): 641-647.

Govender, P., McFarlane, S. A., & Rutherford, R. S. (2010). Fusarium species causing pokkah boeng and their effect on Eldana saccharina walker (lepidoptera: pyralidae). In *Proc S Afr Sug Technol Ass* (Vol. 83, pp. 267-270).

Govender, P., McFarlane, S.A., Rutherford, R.S. (2010). Fusarium species causing pokkah boeng and their effect on *Eldana saccharina* walker (Lepidoptera: pyralidae). *Proc S Afr Sug Technol Ass* 83: 267-270.

Gray, S., Cilia, M. & Ghanim, M. (2014). Circulative, "nonpropagative" virus transmission: an orchestra of virus-, insect-, and plant-derived instruments. *Adv. Virus Res.* 89:141-199.

Gutierrez, S., Michalakis, Y., Van Munster, M. & Blanc, S. (2013). Plant feeding by insect vectors can affect life cycle, population genetics and evolution of plant viruses. *Funct. Ecol.* 27:610-622.

Hall, D.G. (1987). The sugarcane aphid, Melanaphis sacchari (Zehntner), in Florida. *J. Am. Soc. Sugar Cane Technol.* 7:26-29

Hall, D.G. (1988). Insects and mites associated with sugarcane in Florida. *Fla. Entomol.* 71:138-150.

Hilton, A., Zhang, H., Yu, W., & Shim, W.B. (2017). Identification and characterization of pathogenic and endophytic fungal species associated with Pokkah Boeng disease of sugarcane. *The Plant Pathol J.* 33(3): 238.

Hossain, M. I., Ahmad, K., Siddiqui, Y., Saad N., Rahman, M. Z., Harina, A.O., & Benzo, S. K. (2020). Current and prospective strategies on detecting and managing *Colletotrichum falcatum* causing red rot of sugarcane. *Agronomy* 10(9):1253.

Hughes, C.G. & Robinson, P.E. (1961). Downy mildew disease. In: Sugar-cane diseases of the world. Eds: Martin, J.P., Abbott, E.V. and Hughes, C.G. Elsevier, Amsterdam

Hussnain, S. Z., Haque, M. I., Mughal, S. M., Shah, K. N., Irfan, A., Afghan, S., Shahazad, A., Batool, A., Khanum, P., Hussain, K., Nawaz, K., Hassan, M. N., & Hafeez, F.Y. (2011). Isolation and biochemical characterizations of the bacteria (*Acidovorax avenae* subsp. *avenae*) associated with red stripe disease of sugarcane. *Afr J Bioteh*, 10:7191-97.

Hussnain, Z., & Afghan, S. (2006). Impact of major cane diseases on sugarcane yield and sugar recovery. Annual Report, Shakarganj Sugar Research Institute, Jhang, Pakistan.

Jepson, S.B. (2008). Philippine downy mildew of corn. USA: Oregon State University (OSU)-Extension Office.

Johnson S.S. & Tyagi, A.P. (2010). Effect of Ratoon Stunting Disease of Sugarcane yield in Fiji. *S P JNat Appl Sci.* 28: 69-73.

Joshi, N.C. (1954). Fighting wilt diseases of sugarcane in Uttar Pradesh (India). *Sci Cult*, 20: 301-302.

Kirtikar, Singh, G. P. & Shukla, R. (1972). Role of seed material in carryover of wilt disease of sugarcane. *Indian Sugar*, 22: 89-90.

Kumar, B., Yonzone, R., & Kaur, R. (2014). Present status of bacterial top rot disease of sugarcane in Indian Punjab. *Plant Dis Res*, 29(1):68-70.

Kuo, T.T., Chien, M.M. and Li, H.W. (1969). Ethyl acetate produced by *Ceratocystis paradoxa* and *C. adiposum* and its role in the germination of sugarcane buds. *Cand. Jour. Bot.* 47: 1459-1463.

Leu, L.S. & Egan, B.T. (1989). Downy mildew. In: Diseases of sugarcane: major diseases. Eds: Ricaud, C., Egan, B.T., Gillaspie, A.G. and Hughes, C.G. Elsevier, Amsterdam.

Li, J., Zhang, R., Shan, H., Wang, X., Cang, X., Wang, C., et al & Huang, Y. (2020). The Epidemic Characteristics and Control Strategies of Sugarcane Red Rot. *Agric Biotechnol*, 9(3), 124-127.

Lin, Z., Xu, S., Que, Y., Wang, J., Comstock, J. C., Wei, J., et al & Zhang, M. (2014). Species-specific detection and identification of Fusarium species complex, the causal agent of sugarcane pokkah boeng in China. *PloS one*, 9(8), e104195.

Lin, Z., Xu, S., Que, Y., Wang, J., Comstock, J.C., Wei, J., et al. (2014). Species-Specific detection and identification of *Fusarium* Species complex, the causal agent of Sugarcane Pokkah Boeng in China. *PLoS ONE* 9(8): e104195. https://doi.org/10.1371/journal.pone.0104195.

Lockhart, B.E.L, Ireg, H.L.J., Comstock, J.C. (1996). Sugarcane bacilliform virus, sugarcane mild mosaic virus, in sugarcane yellow leaf syndrome. Sirgcircrriie Gerinplasrn Consenation and Exchange: Pi.owedirig.s Aiistrulicin Centre jbi. *Int Agric Res* 67: 108-112.

Lyon, H. L. (1922). A leaf disease of the Tip cane *Proc Hawaiian Sugar Planters Association*: 246.

Madan, V.K., Singh, K., Shukla, U.S. & Saxena, Y.R. (1986). Biochemical studies on sugarcane affected with ratoon stunting disease. *Indian Phytopathol.* **39**: 506-508.

Man V.T. (2007). Plant disease. Hanoi University of Agriculture, Pp 100-103 (in Vietnamese).

Marcone, C. (2002). Phytoplasma diseases of sugarcane. *Sugar Tech*, 4(3): 79-85.

Marcone, C., Schneider, B., & Seemuller, E. (2004). 'Candidatus phytoplasma cynodontis', the phytoplasma associated with Bermuda grass white leaf disease, *Inter J. Syst. Evolut. Microbiol.* 54: 1077-82.

Martin J.P., Hong, H.L., Wismar, C.A. (1961). Pokkah boeng. In: Sugarcane diseases of the world. Vol. 1, Elsevier Publ. Co. New York 542. 247-257

Martin, J. P. (1938). Sugarcane diseases in Hawaii *Advertiser Publication Corporation*, Honolulu Hawaii pp.295.

Martin, J. P., & Wismer, C. A. (1961). Red stripe In: J.P. Martin, E.V. Abbott and C.G. Hughes (Editors), *SugarcaneDiseases of the world Vol. I. Elsevier Publication, CO. Amsterdom*.pp.109- 126.

Martin, J. P., Handojo, H. & Wismer, C. A. (1961). "Pokkah boeng." *Diseases of sugarcane: Major diseases* (1989): 157-165.

Matsumoto, T. (1952). Monograph of sugarcane diseases in Taiwan. Taipei, Taiwan, 61 pp.

Mayeux, M.M., Cochran, J.B. & Steib, R.J. (1998). An Aerated Steam System for controlling Ratoon Stunting Disease. *Transactions of the ASABE* 22: 653-656.

McCoy, R.E., Candwell, A., Chang, C.J., Chen, T.A., Chiyknwski, L.N., Cousin, M.T., Dale, J.L., de Leenw, G.T.N., Golino, D.A., Hackett, K.J., Kirkpatrick, B.C., Marwitz, R., Petzold, H., Sinha, R.C., Sugiura, M., Whitcomb, R.F., Yang, I.L., Zhu, B.M., & Seemiiller, E. (1989). Plant diseases associated with mycoplasmalike organisms. In, The Mycoplasmas, Vol V, Whitcomb, R.F., & Tully, J.G. (Eds), San Diego, USA, Academic Press, pp 545-540.

Mensi, I., Vernerey, M.S., Gargani, D., Nicole, M., & Rott. P. (2014). "Breaking dogmas: The plant vascular pathogen *Xanthomonas albilineans* is able to invade non-vascular tissues despite its reduced genome". *Open Biology* 4:130116. https://doi.org/10.1098/rsob.130116.

Moonan, F. & Mirkov, T.E. (2002). Analyses of genotypic diversity among North, South, and Central American isolates of Sugarcane yellow leaf virus: evidence for Colombian origins and for intraspecific spatial phylogenetic variation. *J. Virol.* 76:1339-1348.

Moonan, F., Molina, J. & Mirkov, T.E. (2000). Sugarcane yellow leaf virus: an emerging virus that has evolved by recombination between luteoviral and poleroviral ancestors. *Virology* 269: 156–171.

Moutia, Y. & Saumtally, S. (1999). Detection from soil and distribution of Ceratocystis paradoxa Moreau, causal agent of pineapple. *AMAS Food Agric Res C, Reduit,* Mauritius, pp 75-82.

Nasare, K., Yadav, A., Singh, A.K., Shivasharanappa, K.B., Nerkar, Y.S., & Reddy, V.S. (2007). Molecular and symptom analysis reveal the presence of new phytoplasmas associated with sugarcane grassy shoot disease in India, *Plant Dis.* 91: 1413-1418.

Nordahliawate, M.S., Nur Ain Izzati, M.Z., Azmi, A.R. & Salleh, B. (2008). Distribution, morphological characterization and pathogenicity of Fusarium sacchari associated with pokkah boeng disease of sugarcane in Peninsula Malaysia. *Pertanika J Tropic Agric Sci* 31: 279-286.

Okabe, N. (1933). Bacterial diseases of plants occurring in Formosa II *Journal Society of Tropical Agriculture* 5: 26-36.

Orian, G. (1956). Occurrence of disease similar to red stripe of sugarcane in Mauritius *Procceding, ISSCT* 9: 1042-48.

Osman, M.A.M. (2021). Isolation and identification of the causal pathogen associated with Pokkah Boeng disease on sugarcane in Upper Egypt. *SVU-International Journal of Agricultural Sciences.* 3(3): 30-39.

Palaniswami, C., Viswanathan, R., Bhaskaran, A., Rakkiyappan, P. & Gopalasundaram, P. (2014). Mapping sugarcane yellow leaf disease affected area using remote sensing technique. *J. Sugarcane Res.* 4:55-61.

Parmessur, Y., Aljanabi, S., Saumtally, S. & Saumtally, A. D. (2002). Sugarcane yellow leaf virus and sugarcane yellows phytoplasma: elimination by tissue culture. *Plant Pathol.* 51:561-566.

Patil, A.S, Hapase, D.G. (1987). Studies on Pokkah boeng disease in Maharashtra. *Ind Phytopath* 40: 290.

Patil, A.S., Singh, H., Sharma, S.R., Rao, G.P. (2007). Morphology and pathogenicity of isolates of *Fusarium moniliformae* causing Pokkah boeng of sugarcane in Maharastra. Ram, R.C., Singh, A (eds.). Microbial Diversity: Modern Trends, Daya Publishers, New Delhi 234-263.

Patro, T., Rao, V.N., & Gopalakrishnan, J. (2006). Association of *Acinetobacter baumannii* with a top rot phase of sugarcane red stripe disease in India *Indian Phytopath*, 59: 501-02.

Pinheiro, P.V., Kliot, A., Ghanim, M. & Cilia, M. (2015). Is there a role for symbiotic bacteria in plant virus transmission by insects? *Curr. Opin. Insect Sci.* 8:69-78.

Rafay, S.A., & Singh, V. B. (1957). A new strain of *Glomerella tucumanensis*. *Curr Sci*, 26:19-20.

Rana, O.S., & Shukla, R. (1968). Ratoon stunting disease of sugarcane and its control in Uttar Pradesh. *Indian Sugar* 18: 541-45.

Rangaswami, G., & Rajagopalan, S. (1973).Bacterial Plant Pathology, Tamil Nadu Agriculture University, Coimbatore.

Rao, G. P. (2004). Sugarcane Pathology. Oxford and IBH Publishing Company, UK 2.

Rao, G.P. & Dhumal, K.N. (2002). Grassy shoot disease of sugarcane. In: Singh SB, Rao GP, Easwaramoorthy S, eds. Sugarcane Crop Management. Houston, USA: Sci Tech Publishing LLC 208–222.

Rao, G.P. & Singh G.P. (2000). Serological diagnosis of Ratoon Stunting Disease of Sugarcane in India. *Sugar Tech*. 2: 35-36.

Rao, G.P., Srivastava, S., Gupta, P., Sharma, S., Singh, A., Singh, S., Singh, M., & Marcone, C. (2008). Detection of sugarcane grassy shoot phytoplasma infecting sugarcane in India and its phylogenetic relationships to closely related phytoplasmas. *Sugar Tech*. 10: 74-80.

Ricaud, C. & Ryan, C.C. (1989). Leaf scald. In: Ricaud, C., Egan, B.T., Gillaspie Jr., A.G. and Hughes, C.G. (eds). Diseases of Sugarcane, Major Diseases, 39–58. Elsevier, Amsterdam, The Netherlands.

Ricaud, C., Egan, B. T, Gullaspie, Jr. A. G.,& Hughes, C. G. (1989). Diseases of Sugarcane-Major diseases. *Elsevier, New York* pp. 399.

Rishi, N., & Chen, C.T. (1989). Grassy shoot and white leaf diseases. Diseases of sugarcane Major diseases, Elsevier Science Publisher, 289-300.

Rochow, W.F. (1982). Dependent transmission by aphids of Barley yellow dwarf luteoviruses from mixed infections. *Phytopathol.* 72:302-305.

Saksena, P., Vishwakarma, S. K., Tiwari, A. K., Singh, A., & Kumar, A. (2013). Pathological and molecular variation in *Colletotrichum falcatum* Went isolates causing Red rot of sugarcane in the Northwest zone of India. *J Plant Prot Res,* 53: 37-41.

Sardana, H. R., Singh, N., & Tripathi, B. K. (2000). Investigation on the relationship between root borer and wilt disease of sugarcane. *Ind J Ento* 62: 11-17.

Sarosh, Mishra, S.R., Singh, K., & Singh, K. (1986). Anatomical studies on sugarcane affected with grassy shoot disease. *Indian Phytopath.* 39: 499- 502.

Satyagopal, K., Sushil, S.N., Jeyakumar, P., Shankar, G., Sharma, O.P., Boina, D., Sain, S.K., Chattopadhyay, D., Asre, R., Kapoor, K.S., Arya, S., Kumar, S., Patni, C.S., Chattopadhyay, C., Pandey, A., Pachori, R., Thakare, A.Y., Basavanagoud, K., Halepyati, A.S., Patil, M.B. and Sreenivas, A.G. (2014). AESA based IPM package for Mustard. National Institute of Plant Health Management, Hyderabad, India. 49 pp.

Satyavir. (2003). Red rot of sugarcane current scenario. *Indian Phytopath,* 56: 245–254.

Schenck, S. & Lehrer, A.T. (2000). Factors affecting the transmission and spread of Sugarcane yellow leaf virus. *Plant Dis.* 84:1085-1088.

Schenck, S. (1990). Yellow leaf syndrome – a new sugarcane disease. Hawaiian Sugar Planters Association: Annual Report 38-39.

Semangun, H. (1992). Host index of plant diseases in Indonesia. Yogyakarta: Gadjah Mada University Press.

Shan, H., Li, W., Huang, Y. *et al.* (2017). First detection of sugarcane red stripe caused by *Acidovorax avenae* subsp. *avenae* in Yuanjiang, Yunnan, China. *Trop Plant Pathol,* 42:137–141.

Sharma, G., Singh, J., Arya, A., & Sharma, S. R. (2017). Biology and management of sugarcane red rot: A review. *Plant Arch,* 17(2):775-784.

Sharma, R., & Tamta, S. (2015). A review on Red rot: The "Cancer" of Sugarcane. *J Plant Pathol Microbiol*, S1: 003. doi:10.4172/2157-7471.S1-003.

Singh, G.R. (1969). Local lesion host of grassy shoot virus of sugarcane. *Curr. Sci.* 38: 148.

Singh, J. & Singh, R. (2015). Sugarcane seed production: Indian scenario. J. *Biotechnol. Crop Sci.* 4:43-55.

Singh, K., Budhraja, T. R., & Agnihotri, V. P. (1977). Survival of *Colletotrichum falcatum* in soil, its portals of entry and role of inoculum density in causing infection. *Int Sugar J*, 79: 43-44.

Singh, V., Baitha, A., & Sinha, O.K. (2002). Transmission of grassy shoot disease of sugarcane by a leaf hopper (Deltocephalus vulgaris Dash & Viraktamath), *Indian J. Sugarcane Technol.* 17: 60-63.

Smith, D.R. & Renfro, B.L. (1999). Pg. 26-28. In: D.G. White (ed.) Compendium of Corn Diseases, 3rd ed. APS Press, St. Paul, MN. 78 pp.

Smith, G.R., Borg, Z., Lockhart, B.E.L., Braithwaite, K.S. & Gibbs, M. (2000). Sugarcane yellow leaf virus: a novel member of the Luteoviridae that probably arose by inter-species recombination. *J. Gen. Virol.* 81: 1865–1869.

Spaull, V.W. & Bailey, R.A. (1993). Combined effect of Nematodes and Ratoon stunting disease on Sugarcane. *Proc South African Sugar Technol Assoc.* 29-33.

Srinivasan, K. V. (1964). Some observations on sugarcane wilt. *J Indian Bot Soc*, 43: 397- 408.

Srinivasan, K. V. and Rao, J. T. (1968). Heat treatment of sugarcane seed material. *Indian Sugar*, 18:683-690.

Srivastava, S., Singh, V., Gupta, P.S., Sinha, O.K., & Baitha, A. (2006). Nested PCR assay for detection of sugarcane grassy shoot phytoplasma in the leafhopper vector *Deltocephalus vulgaris* a first report. *Pl. Pathol.* 55: 25-28.

Subhani, N. M., Chaudhry, A. M., Khaliq, A., & Muhammad F. (2008). Efficacy of various fungicides against sugarcane Red rot (*Colletotrichum falcatum*). *Int J Agric Biol*, 10(6): 725-727.

Suma, S. & Magarey, R.C. (2000). Downy mildew. In: A guide to sugarcane diseases. Eds: Rott, P., Bailey, R.A., Comstock, J.C., Croft, B.J. & Saumtally, A.S. CIRAD & ISSCT.

Suman, A., Lal, S. S., Shasany, A. K., Gaur, A., & Singh P. (2005). Molecular assessment of diversity among patho types of *Colletotrichum Falcatum* prevalent in sub- tropical Indian sugarcane. *World J Microbiol Biotechnol*, 21: 1135-1140.

The 23rd Congress of the International Society of Sugarcane Technologists, New Delhi, India, 1999, 373–81.

Tran-Nguyen, L., Blanche, K.R., Egan, B., & Gibb, K.S. (2000). Diversity of phytoplasmas in northern Australian sugarcane and other grasses. *Pl. Pathol.* 49: 666-679.

Tryon, H. (1923). Top rot of sugarcane *BulletinBur Sugar Experiment Station Division of Pathology. I.*

Vasudeva, R.S. (1955). Stunting of sugarcane. Commonwealth Phytopathological Society, New Delhi.

Vasudeva, R.S. (1960). Report of the Division of Mycology and Plant Pathology, *Sci. Rep Agric. Res. Inst.*, New Delhi, pp 11-130.

Vesminsh, G. E., Chinea, A., & Canada, A. (1978). Causes de la propagacion y desarrollo en Cuba de la enfermedadnayarojabacteriana de la cana de azucar. *Cienc Agriculture* 2: 53-64.

Vishwakarma, S. K., Kumar, P., Nigam, A., Singh, A., & Kumar, A. (2013). Pokkah boeng: An emerging disease of sugarcane. *J Plant Pathol Microbiol*, 4(170), 2.

Vishwakarma, S.K., Kumar, P., Nigam, A., Singh, A. & Kumar, A. (2013) Pokkah Boeng: An Emerging Disease of Sugarcane. *J Plant Pathol Microb.* 4: 170. doi:10.4172/2157-7471.1000170.

Vishwanathan, R., & Malathi, P. (2019). Biocontrol strategies to manage fungal diseases in sugarcane. *Sugar Tech*, 21:202-212.

Vishwnathan, V. (2001). Growing severity of ratoon stunting disease of sugarcane in India. *Sugar Tech.* 3: 154-159.

Viswanathan, R. (2002). Sugarcane yellow leaf syndrome in India: Incidence and effect on yield parameters. *Sugar Cane International.* 17-23.

Viswanathan, R. (2013). Status of sugarcane wilt: One hundred years after its occurrence in India. *J Sugarcane Res,* 3 (2): 86-106.

Viswanathan, R. (2018). Changing scenario of sugarcane diseases in India since introduction of hybrid cane varieties: path travelled for a century. *J Sugarcane Res,*8(1):177–211.

Viswanathan, R., & Padmanaban, P. (2008). Hand book on sugarcane diseases and their management. Sugarcane Breeding Institute, Coimbatore, India.

Viswanathan, R., & Rao, G. P. (2011). Disease scenario and management of major sugarcane diseases in India. *Sugar Tech* 13:336–353.

Viswanathan, R., & Samiyappan, R. (1999). Red rot disease of sugarcane: major constraint for Indian sugar industry. *Sugar Cane,* 5:9–15.

Viswanathan, R., (2012). Sugarcane diseases and their management, Sugarcane Breeding Institute, Coimbatore, India.

Viswanathan, R., Balaji, C. G., Selvakumar, R., Malathi, P., Ramesh, S. A., Annadurai, A., Pazhany, A. S., Manivannan, K., & Nithyanantham, R. (2019). Identification of sources of resistance to wilt caused by *Fusarium sacchari* in Indian sugarcane parental population. *Intern Sugar J,*121:838–846.

Viswanathan, R., Balamuralikrishnan, M. & Karuppaiah, R. (2006). Yellow leaf disease of sugarcane: Occurrence and impact of infected setts on disease severity and yield. *Proc. Sugar Technologists' Assoc. India* 67: 74–89.

Viswanathan, R., Jayanthi, R. & Sanakaranarayanan, C. (2017). Integrated disease and pest management in sugarcane. *Indian Farm.* 67:28-32.

Viswanathan, R., Malathi, P. & Neelamathi, D. (2018). Enhancing sugarcane yield per hectare through improved virus-free seed nursery programme. *ICAR News* 24:4-5.

Viswanathan, R., Malathi, P., Annadurai, A., Naveen, P. C. & Scindiya, M. (2014). Sudden occurrence of wilt and *Pokkah Boeng* in sugarcane and status of resistance in the parental clones in national hybridization garden to these diseases. *J Sugarcane Res,*4(1):62–81.

Viswanathan, R., Malathi, P., Ramesh, S. A., Poongothai, M., & Singh, N. (2006). Current status of sugarcane wilt in India. *Sugar Cane Intern,* 24(4):1–7.

Viswanathan, R., Mohanraj, D., Padmanaban P., & Alexander, K. C. (1994). Possible role of red rot pigments in host-pathogen interaction in sugarcane with red rot pathogen. *Indian Phytopath,* 47: 281.

Viswanathan, R., Poongothai, M., Malathi, P., & Prasanth, C. N. (2015). Sugarcane wilt: simulation of pathogenicity through different methods and environments. *Intern Sugar J,* 117:286–293.

Went, F.A.F.C. (1896). Notes on sugarcane diseases. *Ann. Bot.* 10: 583-600.

Weston, W.H. Jr. (1920). Philippine Downy Mildew of Maize. *J. Agric. Res.* 97-122.

White, W.H., Reagan, T.E. & Hall, D.G. (2001). *Melanaphis sacchari* (Homoptera: Aphididae), a sugarcane pest new to Louisiana. *Fla. Entomol.* 84: 435.

XU ing-Sheng, XU Li-Ping, QUE You-Xiong, GAO San-Ji. & Chen Ru-Kai. (2008). Advances in the Ratoon Stunting Disease of Sugarcane. DOI CNKI:SUN:RYZB.0.2008-02- 017.

Zhang, M., & Jeyakumar, J. M. J. (2018). Fusarium Species Complex Causing Pokkah Boeng in China. In *Fusarium–Plant Diseases, Pathogen Diversity, Genetic Diversity, Resistance and Molecular Markers.* Intech Open.

Chapter - 15

Diseases of Ginger (*Zingiber officinale* Roscoe) and their Integrated Management

Durga Prasad Awasthi and Amar Bahadur

Department of Plant Pathology,College of Agriculture, Tripura, Lembucherra, West Tripura – 799 210

Ginger (*Zingiber officinale* Roscoe) belongs to the family *Zingiberaceae*. The rhizome of the plant is used every day for culinary purposes. It is a household spice crop in India. Since ancient times, it has been recommended as a medicinal plant for many health-related ailments like cold, cough, pain etc. Countries namely China, India, Japan, Taiwan, Nepal, Sri Lanka and Thailand are major producers of Ginger in the world. India is the largest producer of Ginger (Ravindran *et. al*, 2005). The plant is suffered from many biotic agents like fungi, bacteria and viruses etc. However, some major diseases of ginger are rhizome rot or soft rot, bacterial wilt and leaf spot.

1.   Rhizome rot or soft rot of Ginger

It is one of the most important diseases of ginger resulting in heavy yield loss ranging between 4-100% (Nepali *et al.*, 2000). In the year 1907 E J Butler recorded the incidence of this disease in the Surat district of Gujrat, India.

Causal organism (s)

This disease is primarily caused by different species of *Pythium* like *Pythium aphanidermatum*, *P.vexans*, *Pythium myriotylum along* with *Fusarium oxysporum*, *Sclerotium rolfsii* and *Ralstonia solanacearum* etc. *P. vexans* de Bary was reported in

India and Fiji (Ramakrishnan, 1949). Le *et al.*, (2017) reported use of PCR-RFLP is sensitive to detect soft rot caused by *P. myriotylum* strains from artificially infected ginger without the need for isolation for pure cultures.

Symptoms

Pseudo stem, rhizome and collar region of the plants are being infected. Diseased plants appear soft, brown, and water-soaked which gradually decays (Dohroo *et al.*, 2005). Lesions enlarge, plant shows yellowing, drooping and drying of leaves. Initially, the leaves show slight pale colour. The tip of the blades shows yellowing of leaves spreading downward leading to drooping and hanging down of leave and death of infected plants. Basal part of the plant exhibits pale translucent colour which soon turns to water soaked and soft. Rotting may extend from collar regions to rhizomes or upward.

The affected part of the rhizomes may form a watery mass of putrefying tissues enclosed by the tough rind of the rhizome. Rhizomes may show wet rot and dry rot based on the type of infection unlike bacterial infection, in soft rot rhizomes do not produces foul order. Roots also become soft and start rotting.

Epidemiology

It is a seed as well as a soil-borne disease. The fungi may survive by forming resting structures like chlamydospores, oospores etc. Temperatures above 30 º C and soil moisture favour the development of diseases. Water stagnation helps the further spread of disease. Young plants are more susceptible to the pathogen. Infestation of nematodes favours the development of diseases. *P. myriotylum* is found to cause heavy damage in warm humid climates (Dake and Edison, 1989). As nematodes make injury to affected part by stylets thereby helps easy entry of fungi. Soft rot incidence is found to be reduced in soil with low pH and organic carbon >2.25 % (Sharma *et. al.*, 2012).

Management

» Select healthy rhizomes, free from disease-free plants.

» Follow soil solarization which helps to reduce soil-borne inoculum.

» Soil amendments with organic matter and bio-control agents help to reduce soil borne inoculum.

»   Seed treatment and application of *Trichoderma* spp. is found suitable for effective biological management. The growth of *P. aphanidermatum* is found to be inhibited by soaking rhizomes with spore suspension or coated with bioagents namely, *T. viride* and *T. hamatum* (Bhardwaj *et. al.*,1988).

»   Application of *T. harzianum* with Farm Yard Manure and periodic drenching of Copper oxychloride @ 0.3% during the rainy season under field conditions is found to control soft rot disease in ginger (Dohroo *et al.*, 2012).

Treat the seed rhizome before storage and sowing. Fungicides like Copper oxychloride 50%, metalaxyl 8% + mancozeb 64% are found to be effective against the soft rot of ginger.

2. Bacterial Wilt

Bacterial wilt is a widespread and lethal bacterial disease of ginger plants (Lando, 2002). It is a predominant disease in all major ginger-growing states of India. The disease is severe in hot and humid regions where temperature varies between 28°C and 30°C, it also occurs in cold high-altitude areas of the eastern Himalayas. Ginger cultivation in northern and eastern districts has been severely affected by bacterial wilt. The bacterial wilt of ginger imparts serious economic losses and is widely distributed in tropical and subtropical regions of the world (Kumar and Sarma, 2005).

Causal organism (s)

The disease is caused by bacterial wilt pathogen *Ralstonia solanacearum* biovar III (Smith) Yabuuchi. Restriction Fragment Length Polymorphism (RFLP) studies show that there are two major geographical origins of the strains, biovar I and biovar II of American origin. Asian origin consists of biovar III, biovar IV and biovar V correlating to the finding of (Cook et al., 1989). Biovar III causes wilting in ginger within 5 to 7 days after artificial stem inoculation and 7 to 10 days under field conditions in the agro-climatic condition of India (Kumar and Sarma, 2004). It is aerobic, gram-negative, non-capsulate, non-spore-forming, nitrate-reducing bacteria. Rich (1983) reported them as irregular, small, smooth, wet and shiny.

Symptoms

The ginger plant shows water-soaked spots on the pseudo stem which may progress upwards or downwards. Leaves appearing golden yellow is the initial characteristic symptom of bacterial wilt in ginger. Lower leaves show mild drooping and curling of leaves. In the later stage, plants show severe yellowing and wilting symptoms.

Bacterium moves up through the vascular system and blocks water transportation, causing wilting (Tahat and Sijam, 2010). Transversely cut pseudo stems show dark streaks. Infected pseudo stem and rhizomes may extrude ooze which appears milky in nature.

Epidemiology

Bacterial wilt is both soil and seed-borne and mainly appears during monsoon seasons in India. Infected seed rhizomes and water are modes of dissemination for the bacterium. The disease is favoured by a warm climate. The intercultural operation may injure the plant and helps the bacterium infect a new host. Apart from that infestation with nematodes also helps bacteria to infect and cause bacterial wilt of ginger.

Management

1. Use healthy rhizomes free from infection.

2. Rhizome solarization on ginger seeds for 2 to 4 hours reduces bacterial wilt (Kumar and Sood., 2005)

3. Crop rotation with a non-susceptible crop can opted as a management strategy against bacterial wilt.

4. Follow soil amendments with organic matter.

5. Follow good hygienic conditions. Sanitize instruments prior to and during intercultural operations.

6. Bacterial bio-agent like *Pseudomonas fluorescens* is antagonistic to soil-borne pathogens which produce antimicrobial substances, competition for space, nutrients and indirectly through induction of systemic resistance (Kavitha and Umesha, 2007). Thus, the incorporation of bio-agents needs to be encouraged to reduce disease incidence.

Treat the seeds with streptocycline 200 ppm for 30 minutes and dry them in shade before planting. At the onset of disease, beds should be drenched with Bordeaux mixture @ 1% or Copper oxychloride @ 0.2 %.

3. Leaf spot

It is one of the most threatening foliar diseases and was first time reported in India by Ramakrishnan (1941) from Godavari and Malabar regions.

Causal organism (s)

The leaf spot is caused by *Phyllosticta zingiberi*. The pathogen produces pycnidia which are globose to subglobose, dark brown in colour. The pycnidia spores are hyaline, oval to bullet shaped and monoguttulate.

Symptoms

The initial stage of the disease is the appearance of the small spindle to oval or elongated spots mainly on younger leaves. These water-soaked spot-on leaves soon turn into white spots surrounded by dark brown margins and a yellow halo. Lesions enlarge and coalesce to form dead necrotic areas reducing the photosynthetic area of leaves.

Epidemiology

Disease is favoured by wind and rain splash. High humidity and temperature are good for the growth of pathogens. Factors like air temperature, relative humidity and rainfall influence disease incidence.

Management

1. Grow crops under partial shade like mandarin orange.

2. Intercrop of ginger plants with coconut is recommended against leaf spots.

3. Grow resistant varieties. Senapati *et al.,* (2005) reported that PGS-16, PGS-17 and Anamika are moderately resistant to the disease.

4. If disease severity is high spray Copper oxychloride @ 0.25 % or Mancozeb @ 0.2%.

5. Other minor diseases of ginger are mosaic, chlorotic flecks, wet rot and storage rots.

References

Bhardwaj, S.S., Gupta, P.K., Dohroo, N.P. and Shyam, K.R. (1988). Biological control of rhizome rot of ginger in storage. *Indian Journal of Plant Pathology*. 6:56-58.

Cook, D. (1989). Genetic diversity of detection of probes that specify virulence and the hypersensitive response. *Molecular Plant-Microbe Interactions*. 2(3):113.

Dake, G.N. and Edison, S. (1989). Association of pathogens with rhizome rot of ginger in Kerala. *Indian Phytopathology.* 42(1):116-119.

Dohroo, N.P., Ravindran, P.N. and Nirmal, B.K. (2005). Diseases of ginger. In: Ginger the genus Zingiber, 1st Ed. CRC Press, Boca Raton.

Dohroo, N.P., Sandeep, K. and Neha, A. (2012). Status of soft rot of ginger. Technical Bulletin, Department of Vegetable Science, Dr. YS Parmar University of Horticulture and Forestry, Nuani, Solan.

Kavitha, R. and Umesha, S. (2007). Prevalence of bacterial spot in tomato fields of Karnataka and effect of biological seed treatment on disease incidence. *Crop Protection Journal.* 26(7): 991-997.

Kumar, A. and Sarma, Y.R. (2004). Characterization of *Ralstonia solanacearum* causing bacterial wilt in ginger. *Indian Phytopathol.* 57(1):12–17.

Kumar, P. and Sood, A., K. (2005). An eco-friendly approach for the management of bacterial wilt of tomato. *Plant Dis Res.* 20:55–57.

Lando, L.A. D. (2002). Genetic variation of *Ralstonia solanacearum* (E. F. Smith) Yabuuchi et. al. affecting potatoes (*Solanum tuberosum* L.) in Benguet Province, Philippines. PhD Thesis. University of the Philippines, Los Banos, Laguna. 1- 9.

Le, D. P., Smith, M.K. and Aitken, E. A. B. (2017). Genetic variation in *Pythium myriotylum* based on SNP typing and development of a PCR-RFLP detection of isolates recovered from Pythium soft rot ginger. *Appl Microbiol.* 65(4):319-326.

Nepali, M.B., Prasad, R.B. and Sah, D.N. (2000). Survey of ginger growing areas in Syanja, Palpaulni and Arghaknach districts with special emphasis in rhizome rot disease. *Bulletin Lumley Agricultural Research Station,* Nepal.

Ramakrishnan, T.S. (1949). The occurrence of *Pythium vexans* de Bary in South India. *Indian Phytopatholgy.* 2:27-30.

Ramkrishnan, T. S. (1941). A leaf spot disease of *Zingiber officinale* caused by *Phyllosticta zingiberi* N. SP., *Proceedings of the Indian Academy of Sciences - Section B.* 15(4). 167-171.

Saddler, G. S. (2005). Management of bacterial wilt disease. *APS Press ST.Paul, MN.* 121-132.

Ravindran, P.N., Nirmal Babu, K. and Shiva, K.N. (2005). Botany and Crop Improvement of ginger. In: Ravindran, P.N. and Nirmal, B.K., Eds., Ginger: The Genus Zingiber, CRC Press, New York, 15-85.

Rich, A. E. (1983). Potato Disease. New York Press. 11.

Senapati, A.K. and Ghose, S. (2005). Screening of ginger varieties against rhizome rot disease complex in eastern ghat high land zone of Orissa. *Indian Phytopath.* 58 (4): 437-439.

Sharma, B.R., Dutta, S., Roy, S., Debnath, A. and Roy, M.D. (2010). The effect of soil physicochemical properties on rhizome rot and wilt disease complex incidence of ginger under hill agro-climatic region of West Bengal. *Journal of Plant Pathology.* 26:198-202.

Tahat, M.M. and Sijam, K. (2010). *Ralstonia solanacearum:* the bacterial wilt causal agent. *Asian J Plant Sci.* 9(7):385.

Chapter - 16

Diseases of Tobacco (*Nicotiana* spp.) and their Integrated Management

Amar Bahadur[1] and Pranab Dutta[2]

[1]College of Agriculture,Tripura, Lembucherra, Agartala-799210 (Tripura)
[2]School of crop Protection, College of Post Graduate Studies in Agricultural Sciences, Central Agricultural University, Umiam, Meghalaya,

Tobacco (*Nicotiana spp*) is a herbaceous annual or perennial plant in the family Solanaceae grown for leaves. The tobacco plant has a thick, hairy stem and large, simple leaves that are oval. The tobacco plant produces white, cream, pink or red flowers which grow in large clusters, tubular in appearance. Tobacco is usually grown annual, surviving only one growing season. Tobacco is referred to as Virginia tobacco or cultivated tobacco and originates from South America. Tobacco-dried leaves can be cured and used to produce cigarettes, cigars, snuff and for pesticides.

1. Damping off

Damping off (*Pythium* spp) is a fungal disease, the most common and serious disease in tobacco nurseries causing the death of seedlings; it is a soil-borne pathogen that causes damping off problems in nursery beds and the field soon after transplanting. The pathogen causes a watery, soft rot of the lower stem and root system during cool, wet weather. Several soil-inhabiting fungi *Pythium aphanidermatum, Pythium debaryanum, Phytophthora* sp. and sometimes *Rhizoctonia solani* are also involved.

Symptoms

The pathogen attacks the seedlings at any stage in the nursery. A symptom of the disease is the sudden collapse of young seedlings in patches. Watery soft rot of young seedlings, girdling of hypocotyls and finally toppling and death of seedlings, wet rot are the characteristic symptoms. The pathogen spreads quickly and the seedbed causes an enormous loss of seedlings. High humidity, high soil moisture, cloudiness, temperature below 24⁰C, continuous wet weather and location of the nursery in low-lying areas are favourable factors for the high incidence of damping off disease. The disease is noticed in two phases, *viz.* Pre-emergence and Post-emergence damping-off. *Pre-emergence damping-off*-sprouting seedlings are infected and die before emergence from the soil. *Post-emergence damping-off*-it is the most destructive phase. Water-soaked minute lesions appear on the stems near the soil surface, soon girdling the stem, spreading up and down in the stems and one or two days stem rot leading to toppling over of the seedlings. The two-leaved seedlings disappear due to the wet rotting of stems. Generally occurs in patches which spread quickly. Wet rotting and the sudden collapse of seedlings start in circular patches, causing total loss under wet weather. Under favourable conditions, the entire seedlings were killed within 3 to 4 days.

Pathogen

Damping off is caused by *Pythium aphanidermatum, P. debaryanum.* The fungus produces thick, hyaline, thin-walled, non-septate mycelium. It produces lobed sporangia which germinate to produce vesicles containing zoospores. Zoospores are kidney-shaped and biflagellate; oospores are spherical, light to yellowish brown coloured.

Disease cycle

The pathogen survives in the soil as oospores and chlamydospores. The primary infection is from the soil-borne fungal spores and the secondary spread is through sporangia and zoospores transmitted by wind and irrigation water. Over-crowding seedlings, ill-drained nursery beds, high atmospheric humidity (90-100 per cent) and soil moisture, and temperature between 20⁰C to 24⁰ C favour disease development.

Management

Deep summer ploughing, adequate drainage facility, prepare raised seed beds 15 cm high with channels, seed bed with slow burning farm waste materials such as paddy husk before sowing, avoid over-crowding of seedlings by using seed rate (1 to 1.5g/2.5m2), avoid excess watering of the seedlings. Soil solarization has been found effective in nursery soil. Drench nursery bed with formalin (1:50) at 10 cm depth of soil. Bordeaux mixture @ 0.4%, Fytolan, Blitox @ 0.2% (2 gm/litres of water) should apply after 2 weeks sowing. Spraying ridomil MZ 72 W.P. @ 2 g/litres of water at 20 and 30 days after germination. Ridomil should not be sprayed 20 days after germination. Drench the seed bed with 1 per cent Bordeaux mixture or 0.2 per cent Copper oxychloride two days before sowing. Spray the nursery beds two weeks after sowing with 1 per cent Bordeaux mixture, 0.2 per cent Copper oxychloride, 0.2 per cent Mancozeb and repeat subsequently at 4 days intervals under dry weather and at 2 days under wet cloudy weather. Spray 0.2 per cent Metalaxyl at 10 days intervals beginning from 20 days after germination. Seed coat with *Trichoderma harzianum* and *Pseudomonas fluorescens* has been found to effective, talk based formulation applied in nursery bed before sowing is also reduces damping off. Sees treatment with capton and thiram effective again damping off, combination treatments have been very effective.

2. Blue mold

It was first reported in tobacco-growing areas of Australia during the 1800s (Cooke, 1891). The disease was first reported in the United States in 1921 in Florida and Georgia, *P. tabacina* had been identified in Mexico, Canada, South America, the Caribbean, the Middle East, and many European nations. Blue mold affects *Nicotiana* species including all cultivated tobacco. Several wild types of tobacco are also host to the pathogen.

Symptoms

A blue mold problem in nursery beds and the field, the disease is favoured by cool, wet and humid weather, circular yellow spots develop on the upper surface of the leaf and blue-grey fungus sporulating on the underside of leaves can be seen, spots later turn brown, die and may fall out. Lesions may occur on buds, flowers, and capsules. The fungus can grow systemically through the leaf and into the vascular system, resulting in stunted growth with narrow, mottled leaves of the plant (Lucas 1975; Reuveni *et al.*, 1986). On young plants, new leaves are crinkled and twisted and the bud may be killed. The circular patch of seedlings develops yellow leaves. Small patches of dead or dying seedlings are evidence of disease present, and foliar lesions appear yellow, grey, or bluish characteristic of blue mold (Lucas 1975; Wolf *et al.*, 1934). Diseased leaves often become twisted so that the lower surfaces turn upward. Blue mold can affect plants in the field throughout the growing season. Under favourable weather, blue mold can destroy all leaves at any growth stage.

Causal organism

Tobacco blue mold is a downy mildew disease caused by *Peronospora hyoscyami* f.sp. *tabacina* requires a living host to grow and survive (Orlando *et al.*, 2010). The blue mould pathogen belongs to the Kingdom Chromista (Straminipila), Class Oomycetes, Order Peronosporales and Family Peronosporaceae (Kirk *et al.*, 2001; Dick 2002; Voglmayr 2008). The pathogen is an obligate biotrophic parasite, that produces conidiophores and oospores. The hyaline and lemon-shaped conidia borne on tree-like and dichotomously branched conidiophores and terminate acute apices, emerge through the stomata. It's highly destructive to tobacco seed beds, transplants and fields in humid farming zones.

Disease cycle

Asexual spores sporangia do not produce zoospores; infection by direct germination of sporangia. It is assumed that inoculum introduced via windblown Sporangia dispersed thousands of weather and is a primary source of inoculum. Spores produce and released during the morning hours. *Pathogen* requires cool, wet, and cloudy weather to produce spores, under cool and cloudy weather disease cause the destruction of the crop (Lucas 1975). Blue mold cause epidemics by windblown spores from the infected tobacco crops and spread rapidly under favourable weather conditions (Main 1991). Spores are carried on the wind (Davis and Monahan 1991; Lucas 1980) and by the movement of people infected with tobacco transplants (Aylor 1986). Spores land on the damp surface of the tobacco plant and begin to germinate. The pathogen invades the leaf veins, petioles, and vessels of the stem. Overwintering infected crop debris of oospores (sexual spores) is produced, but their role in the disease cycle is poorly understood. The pathogen is capable of overwintering in infected debris, and the role of oospores in disease is not clearly understood (Ristaino *et al.*, 2007)

Management

The pathogen is highly dependent on a cool, humid environment, and the best strategies for maintaining appropriate plant spacing and nutrition to prevent excess humidity in the canopy. Apply fungicides when the pathogens are most vulnerable. Host plant resistance is an economic and environmentally sustainable method for controlling blue mould disease. Tobacco plants with an N gene (resistance gene) inoculated on the lower leaves with tobacco mosaic virus (TMV) occur systemically acquired resistance, which protects against *P. hyoscyami* f. sp. *tabacina* (McIntyre *et al.*, 1981). Biotechnological tools have identified of tobacco genes that resistance against blue mould (Alexander *et al.*, 1993; Borrás-Hidalgo *et al.*, 2006; Kroumova *et al.*, 2007; Lusso and Kuc 1996; Salt *et al.*, 1986; Schiltz 1974).

3. Black Shank

Black shank diseases caused all types of cultivated tobacco in production areas. *Phytophthora parasitica var. nicotianae* pathogenic to tobacco and cause black shank (Apple 1962; Lucas 1975). *Phytophthora parasitica* Dastur var. *nicotianae* (Breda de Haan) is currently used by most pathologists (Shew and Lucas 1991). The pathogen has four races, race O is predominating. The disease was first described by Van Breda de Haan from Java (Indonesia) in 1896 (Shew and Lucas 1991). The disease was introduced into the United States and was first time observed in southern Georgia around 1915 (Shew and Lucas 1991). Black shank disease occurs worldwide, but the disease is most severe in warmer climates (Shew and Lucas 1991). The pathogen affects the crop at any stage of growth. Its soil-borne pathogen occurs heavily in continuous rainfall, high soil moisture and relative humidity, and frequently occurs during nursery causing leaf blight, and blackening of roots and stems leading to the death of seedlings. In advanced stages dry, dark brown/black lesions develop on the stem near the soil line and expand upward, covering half the length of the stem.

Symptoms

The disease affects tobacco plants at all growth stages of all types of tobacco, and symptoms vary with crop age and weather conditions. Pathogen affects the roots and basal stem of the tobacco plant (Shew and Lucas 1991). Young seedlings are very susceptible and have characteristic "damping-off" during wet and mild weather. During rainy weather, plants get the infection by the pathogen from the soil as splashed onto the leaves. The leaf infections result from zoospores splashing onto the leaf surface. Leaf spots appear as water-soaked and light green lesions rapidly increase turning brown and necrotic lesions up to 8 cm in diameter (Shew and Lucas 1991). Pathogen colonization of root tissues in susceptible varieties ultimately reaches the stem, resulting in the characteristic black shank symptom and plant death. Symptoms of the black shank are characteristic yellowing and wilting of leaves, plant die within a few days (Shew 1987). Root tips and wounds are the primary sites of infection; stem lesions on the stem above 30 cm soil surface. When the affected stem is split open, the pith region is found to be dried up in disc-like plates showing black discolouration. The black shank is more severe in poorly drained areas, high temperatures, humidity, rainfall, and soil moisture lead to disease development.

Causal organism

Black shank disease is caused by *Phytophthora nicotianae* and is widely recognized, belongs to the Kingdom- Straminipila, phylum-Oomycota, class- Oomycetes, order Peronosporales and family Pythiaceae. Hyphae of the fungus are hyaline, coenocytic, and typically irregular in width. Sporangia are pear-shaped and have very conspicuous papillae and sporangiophores in a sympodial fashion. Sporangia germination is direct by the production of hyphae and indirect production by zoospores. The zoospore is kidney-shaped with two flagella, the posterior flagellum is whip-like and the anterior tinsel-type, encyst and germinate single germ tube (Hickman 1970). The fungus produces thick-walled chlamydospores, at the terminal or intercalary of hyphae. Chlamydospores serve as the survival as primary inoculums; they can survive for 4 to 6 years in the soil. The pathogen is heterothallic and generally requires two mating types are required for the production of oospores. Oospores are sexual spores, globose smooth and light yellow coloured; antheridium is amphigynous and permanently attached to the oogonium. The fungus can be isolated from soil and plant material by baiting procedures or using a selective agar medium (Shew and Lucas 1991). An excellent selective medium is cornmeal agar, 5% V8 juice with 20g agar.

Disease cycle

The fungus lives as a saprophyte on crop residues and is also present as dormant mycelium in soil, oospores and chlamydospores causing primary infection. In warm, moist soil, chlamydospores germinate and infect tobacco roots by germ tube or produce sporangium and produce 5-to-30 zoospores. Zoospores swim and move toward nutrient around the root tips and wounds on the host plant, Zoospore contacts the root surface, and encysts germinate to form a germ tube that directly penetrates the host epidermis. Sporangia serve as secondary inoculums, root and stem colonization results in typical root rot and black shank symptoms, spread by wind and irrigation water, transport of soil, farm implements and animals (Ferrin and Mitchell 1986). The cycle repeats with the sporangia as a polycyclic disease. Cloudy weather and temperature below 22^0C are favour for the sudden outbreak of the disease. Cardinal temperatures of the life cycle of the pathogen are $10\text{-}12^0$ C minimum, $24\text{-}30^0$ C optimum, and 36^0 C maximum (Shew and Lucas 1991). It is a warm-weather disease; soil temperatures above 20^0 C are required for infection (Jacobi *et al.* 1983). High soil moisture enhances disease development, and sporangium production (Shew 1983, Shew and Lucas 1991). Black shank occurs in acid and alkaline soils, with optimum soil pH for disease development between pH 6 to 7. Levels of calcium and magnesium are positively correlated with black shank severity

(Shew and Lucas 1991). The disease severity increases in the presence of root-knot nematodes and lose their resistance.

Management

Black shank can manage through crop rotation, fungicides, nematode control, cultural practices, and resistant varieties (Shew and Lucas 1991). Stalk and root removal and destruction of affected plants in the field after harvest. Crop rotation is very effective at reducing inoculum density, duration of crop rotation is at least two to four years to eliminate the pathogen from soils. Cover the seed beds with paddy husk at a 15-20 cm thick layer and burn. The widely used to control black shank by planting resistant varieties, the Kanchan variety is tolerant to black shank. Prepare raised seed beds and select disease-free seedlings for transplanting. Drainage the nursery and drench the nursery beds with 0.2% per cent copper oxychloride, two days before sowing. Spray the nursery beds with Mancozeb, Ziram, Metalaxyl, Captafol, and Copper oxychloride at 0.2% or 1% Bordeaux mixture and repeat after 10 days after two weeks of sowing. Spray copper oxychloride (Blitox or Fytolan) @ 2gm/l when seedlings are 50-60 days old. The fumigants chloropicrin and 1,3-dichloropropene reduce nematode populations that enhance diseases and severity. The disease is favoured at greater than 6 pH and suppressed at lower pH by increased activity of aluminium (Al^{3+}), which is highly toxic to *Phytophthora* spp. Soil pH between 5.5 and 6.0 favourable for growing tobacco. Two single genes were incorporated into tobacco cultivars that found complete resistance to race 0, and no resistance to race 1 of the pathogen (Carlson *et al.*, 1997; Johnson *et al.*, 2002; Csinos 2005; Sullivan *et al.*, 2005).

4. Frog eye spot

The disease is found most commonly on lower, more mature leaves of the plant, but can also affect green tissues. Under conducive environmental conditions cause severe damage to leaves yield and quality is also reduced. The disease may occur in the seedlings, leading to the withering of leaves and death of the seedlings, disease is seen 4-5 weeks after germination, spots appear first on basal leaves and gradually spread to the upper leaves of plants. This is a common disease of tobacco, developing in the nursery and field, spots, circular 2-15 mm diameter, brown, grey or tan, with dark borders, resembling frog eye appear on the lower leaves of the seedlings (Lucas 1975).

Symptoms

The disease appears mostly on mature, lower leaves as small ashy grey spots with brown borders. The pathogen infects all stages of the crop; spots are characterized by a dark brown margin with a pale tan/white middle. Initially, the lower leaves show brown, round lesions which resemble a frog-eye shape with a grayish center. The disease spreads upwards, and under favourable conditions, lesions may coalesce to become bigger lesions resulting in the drying of leaves. Yellow halo can be seen on green leaves, out from the dark margin. Dark-coloured fruiting bodies are created, and in the center of the spot, conidia emerge from the stroma. Frequent watering and wet weather lead to high humidity (80-90 per cent) and temperature around 27⁰C Close spacing and excess application of nitrogenous fertilizers are favourable for the disease development.

Pathogen

Frogeye leaf spot is caused by *Cercospora nicotianae* in tobacco; it can also reproduce on several common weeds. Fungus mycelium is intercellular and collects beneath the epidermis and clusters of conidiophores emerge through stomata. The conidiophores are septate, dark brown at the base and lighter towards the top bearing 2-3 conidia. The conidia are hyaline, slender, slightly curved and 2-12 septate. Leaf spot pathogens of tobacco are favoured by warm, humid environments. Asexual reproductive structures called stroma for conidia on conidiophore, conidia are long, multi-septate and brown, after dispersal to susceptible host tissues, conidia germinate to form new lesions.

Disease cycle

The fungus overwinters in the crop residues, plant debris and the collateral host. The primary infection is caused by seed and soil-borne inoculum. The secondary spread is through wind-borne conidia. The conidia are dispersed by water and air. Spores germinate and infect the tissue of the leaf, forming a necrotic lesion and fruiting bodies (stroma) are formed within the leaf lesion and erupt from plant tissues to release spores.

Management

Cultural practices are important to limiting damages by this pathogen, removing and burning plant debris can help reduce disease incidence and severity of frog eye leaf spot, avoid excess nitrogenous fertilization, and row spacing can also reduce incidence and severity. Crop rotation with non-host crops helps in breaking the life cycle of the disease. Regulate irrigation, Spray the crop with 0.4 per cent Bordeaux mixture or Carbendazim 50 WP 3 gm and repeat after 15 days. Azoxystrobin 23% SC @ 0.1% at 30 days and 40 days after germination, effectively controls frog eye spot disease.

5. Brown spot

Brown spot is a major disease of tobacco in India, caused by *Alternaria longipea* and *Alternaria alternata* much prevalent in the field, initially appears on lower and older leaves as small dark brown, circular lesions usually surrounded by a bright yellow halo, which spread to upper leaves, petioles, stalks and capsules. In warm and humidity the leaf spots are 1-3 cm in diameter, centre are necroses, and turn brown with characteristic target board appearance outline. At near maturity, leaf spots are surrounded bright yellow halo, due to the production of toxin 'alternin'. The fungus persists as a mycelium in the dead tissue of tobacco for several months, under favourable weather conidial production starts and infects the lowermost leaves and gradually spreads to the upper leaves by repeating infection cycles of tobacco at any age under high humidity. Dark brown, sunken, elongated spots appear on stems, petioles, seed capsules and stalks. Disease severity is positively correlated with cloudy, wet weather. Removal and destruction of diseased plant debris can reduce primary infection, Excess doses of nitrogen fertilisers should be avoided. High-volume sprayers like knapsack sprayers should be used, Mancozeb @ 20 g in 10 litres of water in combination with streptomycin 3 gm / 10 litres of water sprayed on the crop. Weekly, spraying of fungicides such as maneb, Zineb @ 2kg/ha, or Benomyl or Thiophanate methyle @ 1kg/ha. Depending on the intensity, sprays should be given at 10 days intervals.

6. Powdery mildew

Erysiphe cichoracearum var. *nicotianae* fungus is an ectophytic cause of powdery mildew disease that appears as small, white isolated patches on the upper surface of the leaves. Later, covers the entire lamina. The disease initially appears on the lower leaves, the rest of the leaves are also infected and sometimes powdery growth can be seen on the stem. The severe infection leads to defoliation and a reduction in the quantity and quality of leaves. The fungus produces hyaline, septate and highly branched mycelium, producing short stout conidiophores from the mycelium and conidia in chains. Conidia are barrel-shaped or cylindrical, hyaline. Cleistothecia are black, and spherical with numerous septate appendages. They contain asci, Ascus contains two ascospores that are oval to elliptical single celled. Close planting and excess doses of nitrogenous fertilizers, low temperatures (16-23°C) humid cloudy weather favour disease. The dormant mycelium and cleistothecia in plant debris in soil, cause primary infection from soil and secondary spread by windblown conidia. Remove and destroy the affected leaves, balanced fertilizers, and avoid overcrowding of plants. Spray dinocap at 375 ml or Carbendazim at 500g/ha.

7. Anthracnose

Disease caused by *Colletotrichum tabacum*, and can be seen at any stage of the nursery, light green to white water-soaked lesions develop on young leaves, white areas in the centre and coalesce to form large necrotic lesions, cause a heavy loss in the nursery if favourable conditions prevail, the infection starts on lower leaves as pale-brown circular spots with depressed center and outlined brown margin with slightly raised, Under humid weather condition, dark brown, elongated, sunken necrotic lesions appear on midrib, petiole and stem. The primary infection starts from affected aerial plant parts left in the soil, pathogen persists in the soil on plant debris. The optimum temperature for the disease is 18⁰C, with high relative humidity, and reduced light favourable for the outbreak of this disease. Diseases can manage through raised seed beds and the burning of plant debris, which help in reducing the initial inoculums. Removal and destruction of diseased crop debris minimize the pathogen population in the soil. Rogue diseased seedlings with necrotic lesions on stem, spraying with Bordeaux mixture at 1.0% and Zineb 0.2% at fortnight intervals.

8. Wildfire disease

Disease caused by *Pseudomonas tabaci*, the bacterium is a rod, motile with a single polar flagellum, non-capsulated, non-spore-forming and Gram-negative. Symptoms occur at any stage of plant growth including the nursery seedlings. Wildfire is characterized by dark brown to black spots with a yellow halo that spreads quickly causing withering and drying of leaves, lesions develop on the young stem tissues leading to withering and drying of the seedlings. Numerous water-soaked black spots appear in the fields, later becoming angular and restricted by the veins and veinlets. The angular lesion is brown, dark brown or black, much larger than the wildfire lesion, with little or no chlorotic halo. Several spots coalesce to cause necrotic patches on the leaves and the entire leaf is covered with enlarged spots with yellow haloes. Wildfires and angular leaf spots are favoured by cloudy wet weather, under humid weather conditions, the disease spreads very fast and covers the entire plant blighted appearance. Close planting, humid wet weather and wind favour disease. The bacterium survives in the infected crop residues in the soil and serves as the primary source of infection. The secondary spread by wind splashed rainwater and implements. Solanaceous weeds are hosts of this pathogen. Disease management by removing and burning the infected crop residues in the soil and avoiding very close planting. All seedbed tools, particularly those used for clipping/mowing, should be regularly sterilised with bleach or a copper-based compound. The disease is favor excessive fertility with high N and low K fertilization. Fumigation of seedbeds eliminates initial inoculum in the seedbeds. Seedlings should be sprayed with a combination of Kocide 101, Copper Oxychloride and systemic acquired resistance compounds such as Actiguard and Bion.

9. Tobacco Mosaic Virus

TMV hosts are tobacco, tomato and other solanaceous plants (Jones *et al.*, 1991). This virus disease causes varying degrees of damage depending on the infection. In the early stages of infection, they show stunted growth reducing the yield and quality (Shew and Lucas 1991). The yield losses by TMV in tobacco are estimated at 1% due to the use of resistant varieties; and losses up to 20% are reported in tomatoes, infect tomatoes, distortion of fruits, delayed ripening, fruit colour and reduction in commercial value. The Tobacco Mosaic Virus is the first virus to be discovered and purified. Martinus W. Beijerinck in 1898, Netherlands concepts that TMV was small and infectious, that could not be cultured except in living, growing plants. Wendall Stanley 1946 awarded the Nobel Prize for his isolation of TMV crystals, composed entirely of protein. F.C. Bawden and N. Pirie, in England, same period demonstrated that TMV is a ribonucleoprotein (RNA) and coat protein. TMV is the first virus characterized by X-ray crystallography as a helical structure and the first RNA virus genome that was completely sequenced. Its first virus the resistance gene (N gene) was characterized in the plant (Scholthof 2004). It's a contagious disease that spreads through contact with labour and implements used for intercultural (Scholthof 1999).

Symptoms

The symptoms depend on the host plant, the age of the infection, environmental conditions and the virus strain. Tobacco mosaic virus (TMV) symptoms include mosaic, mottling, necrosis stunting, leaf curling, and yellowing of plant tissues. The early season infected plants are usually stunted, chlorotic, mottle and curled leaves, in severe leaves are narrowed, puckered, thin and malformed, later, under hot weather dark brown necrotic spots develop this is called mosaic burn or scorching. The light discolouration along the veins of the youngest leaves develops a characteristic light and dark green pattern turning into irregular blisters.

Pathogen

TMV is a large group of viruses within the genus *Tobamovirus*. The rod-shaped virus particles measure about 300 nm x 15 nm. A single particle composed of 2,130 coat protein (CP) that envelope the RNA molecule of about 6,400 nucleotides. The virus moves cell-to-cell through plasmodesmata which are connected plant cells and reach the vascular system rapid, systemic spread through the phloem to the roots and tips of the growing plant (Ding 1998, Nelson *et al.*, 1998).TMV enters through the wounds and virus particles expose RNA. The positive-sense serves as messenger RNA (mRNA) and translated host ribosomes, soon proteins have synthesized and replicated with the 3' end of the positive-sense RNA for production of a negative sense. The negative sense RNA template produces full-length genomic positive-sense RNA as well as negative sense sub-genomic RNAs. The sub-genomic RNAs are translated by the host ribosomes to produce movement protein and coat protein. The coat protein then interacts with positive-sense RNA for the assembl of progeny virions. When the cells are broken released to new plants infection, positive-sense RNA wrapped in movement protein and infects adjacent cells (Harrison *et al.*, 1999).

Disease cycle

TMV is made up of a piece of nucleic acid (RNA) and a surrounding protein coat. Once inside the plant cell, the protein coat falls away and the nucleic acid portion in the cell produces more virus nucleic acid and virus protein, disrupting the normal activity of the cell. TMV can multiply only by living cells but survive in a dormant state in dead tissue and can infect growing plants. The virus is viable in plant debris in the soil, serves as a source of inoculums and the longevity of the virus is very high. TMV is easily transmitted when an infected leaf rubs with a healthy plant, by contaminated tools, and workers whose hands contaminated TMV after smoking cigarettes, wounded cell provides a site of entry for TMV. The virus has a wide host range belonging to nine different families. The virus is not seed-transmitted in tobacco, but tomato seeds transmit the virus. Contaminate seed germinating becomes infected. TMV spread from plant to plant through workers' hands, clothing and tools. This is called 'mechanical' transmission. Sucking insects such as aphids do not spread TMV. Chewing insects such as grasshoppers and caterpillars occasionally spread the virus.

Management

Phytosanitary measures are very important for the control of the disease. Transplant virus-free plants, remove all weeds and crop debris and discard infected plants. All tools should be washed with soap or 10% household bleach to inactivate the virus. Washing hands in soapy water before and after field operations, rouging the diseased plants early in the season and avoiding smoking during field operations. Prophylactic sprays virus inhibitors by plant origin like *Basella alba* and *Bougainvillea spectabilis* and neem leaf extracts @ 1% dilution on 30th, 40th and 50th day of planting. Susceptible seedlings should not be transplanted and infected plants should be removed from the field within 3 weeks after planting and replaced with healthy seedlings. Crop rotation practices through resistant plants reduce the number of inoculums in the field. Grow resistant varieties such as TMV RR2, and TMV RR3. TMV can be inactivated by dipping hands in milk before planting. Inoculating mild strains of the virus onto young plants can protect against subsequent infection of severe strains of TMV, called "cross-protection." Transgenic plants alternative strategies for virus control by Genetic engineering techniques, to express the TMV coat protein gene in transgenic plants (Abel *et al.*, 1986). TMV contaminated tobacco seed treatment with a 10% solution of trisodium phosphate for 15 minutes.

10. Leaf curl

Tobacco leaf curl virus infections occur at any stage of plants, plants are stunted, curling of leaves with clearing and thickening of veins; twisting of petioles; puckering of leaves; rugose, brittle and enations are the important symptoms of tobacco leaf curl disease. Three forms of leaf curl are observed; first, the leaf margins curl downward and thickening of veins with enation on the lower surface, second crinkle and curling of the whole leaf edge dorsal side with enation on the veins and the lamina towards the ventral side between the veinlets and third the curling of leaves towards the ventral side with the clearing of the veins and enations absent. Disease caused by tobacco leaf curl geminivirus, non-enveloped, 18 nm diameter circular ss DNA genome, the virus has a narrow host range in eight plant families and is transmitted by white fly (*Bemisia tabaci*). Brinjal, sunflower growing nearby tobacco fields to encourage the build-up of whiteflies. Manage through remove and destroy the infected plants, rogue out alternate weed hosts around nursery area which harbour the virus, spray methyldemeton at 0.1% to control vector. Yellow-sticky traps (20 cm x 15 cm size) 5 nos per acre on iron sheet painted with yellow colour with castor oil. If the population is 100 per each sticky trap, insecticides (Imidacloprid 2.5 ml in 10 litres of water) are sprayed at 10 days intervals commencing from 4 weeks after germination.

11. Root-knot nematode

Root-knot nematodes *Meloidogyne javanica/ Meloidogyne incognita* prevalent in most of the soil and nurseries are prone to disease; adequate soil moisture favours the disease development. Several weeds around nurseries are alternate hosts of root-knot nematodes as the build-up of the population in the soil. Symptoms of root-knot disease are yellowing of leaves, stunted seedlings, wilting of plants, and premature death of seedlings in patches in seed beds. Seedlings show several galls on the roots and vary in size which gives the seedlings a sickly appearance. Soon after the nursery season, ploughing of the nursery area thoroughly destroys the nematode population by exposing them to the hot sun. The area should be ploughed subsequently, 3-4 times during summer months. Before raising the nursery, the soil must be tested for nematodes. Before sowing seed beds, destroy nematode larvae and egg masses by burning paddy husk with slow-burning farm waste material for initial seedlings protection. Soil solarization of nursery bed by covering white polyethene sheet during summer for about 6-8 weeks. Brinjal and tomato grown fields previously, the nursery should not grow. Crop rotation of the nursery field with groundnut, redgram, marigold, cotton, gingelly and chillies for more than three years. The nursery site should be changed every year, if the nursery is infected with nematodes, the nursery should not grow in the next year. Use tray seedlings carrying *Paecilomyces lilacinus* and *Pochonia chlamydosporia*.

12. Broomrape

Orobanche cernua and *Orobanche ramose* are total root parasites affecting the yield and quality of tobacco. Broomrape which lacks chlorophyll, shoots emerge in clusters of 50-100 shoots around the base of a single tobacco plant attached for nutrients and water, resulting in yield loss of 24 to 52%. Infested plants are stunted, and leaves turn pale and wilt. High soil moisture and rain after planting, and low soil temperature during winter months encourage heavy incidence of Orobanche. The young shoot of the parasite emerges from the soil at the base of the plants 5-6 weeks after transplanting. It is an annual, fleshy flowering plant, with a stem 10-15 inches long. The stem is pale yellow or brownish red and covered by small, thin, brown scaly leaves and the base of the stem thickened, white-coloured flowers appear in the leaf axils. The fruits are capsules containing small, black, reticulate and ovoid seeds that remain dormant in the soil for several years. Seeds spread from field to field by irrigation water, animals, human beings and implements. Dormant seeds are stimulated to germinate by the root exudates of tobacco and attach to the roots by forming haustoria, producing shoots and flowers. Orobanche also attacks brinjal, tomato, cauliflower, okra, turnip and cruciferous crops. Deep ploughing 2-3 times in summer and seed to deeper depth in reducing the emergence. Rogue out the tender shoots of the parasite before flowering and seed set. Spray the soil with 25 per cent copper sulphate, 0.1 per cent Allyl alcohol, few drops of kerosene directly on the shoot, Grow trap crops like chilli, moth bean, gingelly, black gram and green gram sorghum, and cowpea to stimulate seed germination and kill the parasite. Avoiding growing brinjal, tomato and bhindi in the sick fields. Orobanche shoots should be destroyed by burning.crop

References

Abel P.P., Nelson R.S., De B., Hoffmann N., Rogers S.G., Fraley R.T and Beachy R.N (1986). Delay of disease development in transgenic plants that express the tobacco mosaic virus coat protein gene. *Science* 232:738-743.

Alexander D., Goodman R.M., Gut-Rella M., Glascock C., Weymann K., Friedrich L., Madoox D., Ahl-goy P., Luntz T., Ward E and Ryals J (1993). Increased tolerance to two oomycete pathogens in transgenic tobacco expressing pathogenesis-relatedprotein1a. *Plant Biol.* 90, 7327–7331.

Apple J.L (1962). Physiological specialization within *Phytophthora parasitica* var. *nicotianae*. *Phytopathology* 52:351-354.

Aylor D.E (1986). A framework for examining inter-regional aerial transport of fungal spores. *Agric. Forest Meteorol.* 38, 263–288.

Borrás-Hidalgo O., Thomma B.P.H.J., Collazo C., Chacón O., Borroto C.J., Ayra C., Portieles R., López, Y and Pujol M (2006). EIL2 transcription factor and glutathione synthetase are required for defense of tobacco against tobacco blue mold. *Mol. Plant–Microbe Interact.* 19, 399–406.

Cooke M.C (1891). Tobacco disease. Gard. Chron. 9, 173.

Carlson S.R., Wolff M.F., Shew H.D and Wernsman E.A (1997). Inheritance of resistance to race 0 of *Phytophthora parasitica* var. *nicotianae* from the flue-cured tobacco cultivar Coker 371-Gold. Plant Disease 81:1269-1274.

Csinos A.S (2005). Relationship of isolate origin to pathogenicity of race 0 and 1 of *Phytophthora parasitica* var. *nicotianae* on tobacco cultivars. Plant Disease 89:332-337.

Csinos A.S (1999). Stem and root resistance to tobacco black shank. Plant Disease 83:777-780.

Erwin D.C and Ribeiro O.K (1996). Phytophthora Diseases Worldwide. *American Phytopathological Society* Press, St. Paul, MN.

Ding B (1998). Intercellular protein trafficking through plasmodesmata. Plant Mol. Biol. 38:279-310.

Dick M.W. (2002). Towards an understanding of the evolution of the downy mildews. In: Advances in Downy Mildew Research (Spencer Phillips, P.T.N., Gisi, U. and Lebeda, A., eds.), pp. 1–57. Dordrecht: Kluwer.

Davis J.M. and Monahan, J.F (1991). Climatology of air parcel trajectories related to the atmospheric transport of *Peronospora tabacina. Plant Dis.* 75, 706–711.

Ferrin D.M and D.J. Mitchell (1986). Influence of initial density and distribution of inoculum on the epidemiology of tobacco black shank. *Phytopathology* 76:1153-1158.

Hickman C.J (1970). Biology of Phytophthora zoospores. *Phytopathology* 60: 1128-1135.

Harrison, B.D and Wilson T.M.A. (1999). Tobacco mosaic virus: Pioneering research for a century. Phil. Transact. Royal Soc. London B. 354: 517-685.

Jacobi W.R., Main C.E., and Powell N.T (1983). Influence of temperature and rainfall on the development of tobacco black shank. *Phytopathology* 73: 139-143.

Johnson E.S., Wolff M.F., Wernsman E.A, Atchley W.R. and Shew H.D (2002). Origin of the black shank resistance gene, Ph, in tobacco cultivar Coker 371-Gold. *Plant Disease* 86:1080-1084.

Jones J.B., J.P. Jones, R.E. Stall and T.A. Zitter (1991). Compendium of Tomato Diseases. APS Press, St. Paul, MN.

Kirk P. M., Cannon P. F., David J. C and Stalpers J.A (2001). Ainsworth & Bisby's Dictionary of the Fungi, 9th edn. Wallingford: CAB International.

Kroumova A.B., Shepherd R.W and Wagner G.J (2007). Impacts of T-phylloplanin gene knockdown and of Helianthus and Datura phylloplanins on *Peronospora tabacina* spore germination and disease potential. *Plant Physiol.* 144, 1843–1851.

Lucas G.B. (1980). The war against blue mold. Science (Washington, DC) 210, 147–153.

Lucas G.B. (1975). Diseases of Tobacco. Raleigh, NC: Biological Consulting Associates.

Lusso M and Kuc J. (1996). The effect of sense and antisense expression of the PR-N gene for b-1,3-glucanase on disease resistance of tobacco to fungi and viruses. *Physiol. Mol. Plant Pathol.* 49, 267–283

Main C.E (1991). Blue mold. In: Compendium of Tobacco Diseases (Shew, H.D. and Lucas, G.B. eds), pp. 5–9. St. Paul, MN: The *American Phytopathological Society.*

McIntyre J.L., Dodds J.A and Hare J.D. (1981). Effects of localized infections of Nicotiana tabacum by Tobacco Mosaic Virus on systemic resistance against diverse pathogens and an insect. *Phytopathology*,71, 297–301.

Nelson R.S and A.J.E. van Bel (1998). The mystery of virus trafficking into, through and out of vascular tissue. *Prog. Bot.* 59:476-533.

Orlando B.H., Bart P. H. J. T., Y. Silva, O. Chacón and M. Pujol (2010). Tobacco blue mould disease caused by *Peronospora hyoscyami* f. sp. *Tabacina. Molecular Plant Pathology* .11(1), 13–18 . DOI: 10.1111/J.1364-3703.2009.00569.X

Reuveni M., Tuzun, S., Cole J.S., Siegel M.R and Kuc, J. (1986). The effects of plant age and leaf position in the susceptibility of tobacco to blue mold caused by *Peronospora tabacina*. *Phytopathology*, 76, 455– 458.

Ristaino J.B., Johnson A., Blanco-Meneses M and Liu B. (2007). Identification of the tobacco blue mold pathogen, *Peronospora tabacina*, by polymerase chain reaction. *Plant Dis.* 91, 685–691.

Salt S. D., Tuzun S. and Kuc, J. (1986). Effect of β-ionone and abscisic acid on the growth of tobacco and resistance to blue mold: mimicry of effects of stem infection by *Peronospora tabacina* Adam. *Physiol. Mol. Plant Pathol.* 28, 287–297.

Schiltz P. (1974). Action inhibit rice de la β-ionone au cours du développement de *Peronospora tabacina. Ann. Tabac.* 11, 207–216.

Shew H.D (1983). Effects of soil matric potential on infection of tobacco by *Phytophthora parasitica* var. *nicotianae. Phytopathology*, 73: 1090-1093.

Shew H.D and Lucas G. B (1991). Compendium of Tobacco Diseases. *Amer. Phytopath. Soc.*, St. Paul, Minnesota.

Scholthof K.B. G (2004). Tobacco mosaic virus: A model system for plant biology. *Annu. Rev. Phytopathol.* 42:13-34

Scholthof K-B.G., J.G. Shaw, and M. Zaitlin (1999). Tobacco mosaic virus: 100 years of contributions to virology. APS Press. St. Paul, MN.

Shew H.D (1987). Effect of host resistance on spread of *Phytophthora parasitica* var. *nicotianae* and the subsequent development of tobacco black shank under field conditions. Phytopathology 77:1090-1093.

Sullivan M.J., T.A. Melton, and H.D. Shew (2005). Managing the race structure of *Phytophthora parasitica* var. *nicotianae* with cultivar rotation. *Plant Disease* 89:1285-1294.

Voglmayr H. (2008) Progress and challenges in systematics of downy mildews and white blister rusts: new insights from genes and morphology. *Eur. J. Plant Pathol.* 122, 3–18.

Wolf F.A., Dixon L.F., McLean R. and Darkis F.R. (1934) Downy mildew of tobacco. *Phytopathology*, 24, 337–363.

Chapter - 17

Diseases of Cotton (*Gossypium* spp.) and their Integrated Management

Amar Bahadur[1] and Pranab Dutta[2]

[1]College of Agriculture, Tripura, Lembucherra, Agartala-799210 (Tripura)
[2]School of crop Protection, College of Post Graduate Studies in Agricultural Sciences, Central Agricultural University, Umiam, Meghalaya,

The cotton crop (*Gossypium* spp*)* is an important industry sector of the economy of agriculture and leads as a cash crop that provides income to farmers and industrialists (Abbas and Ahmad 2018). It is a valuable farming system worldwide (Amin *et al.* 2018; Rahman *et al.* 2018; Tariq *et al.* 2018). The three leading countries in cotton production are China, the United States, and India. Cotton belongs to the Malvaceae family and prime source of fiber worldwide. Cotton grows in tropical and subtropical regions with warm and humid climatic conditions. Cotton crop is suffering from biotic factors that are responsible for crop yield losses. Cotton crop is vulnerable to biotic stresses caused by plant fungal, bacterial, and viral pathogens in production. *Fusarium* and *Verticillium* wilt, *Alternaria* leaf spot and seedling diseases, boll rot, leaf curl disease, and bacterial blight are the major diseases of cotton production and result in poor fiber quality. The soil borne diseases such as root rots caused by *Rhizoctonia solani* and *R. batalicola* and wide spread in the north zone, wilts caused by *Fusarium oxysporum* and *Verticillium dahlia, Fusarium* wilt in the black soils of Gujarat, Madhya Pradesh, Maharashtra, Karnataka and Gujarat and *Verticillium* wilt in Tamil Nadu. Cotton leaf curl emerged and main risk to all cotton-growing areas in viral disease complex, whitefly-transmitted begomoviruses (family Geminiviridae) predominant in South Asia and Africa. The etiology estimates the economic impact of diseases, which helps to develop management strategies. The strategies which contribute to control by chemical control, sowing disease-free seed and resistant varieties, employing crop rotation, and removing infected plant debris along with suitable practices should be part of an integrated disease management strategy. The struggle in management continues against cotton diseases under sustainable agriculture.

1. Seedling Diseases

The seedling disease is a worldwide problem in cotton-growing areas, there are different fungal species involved in cotton seedling diseases and kills the entire seedling population.

Symptoms

The seeds and seedlings are at pre-emergence and post-emergence disease-causing organisms attack and decay of seeds and young seedlings, partial or complete stem girdling stunted growth, and seedling rot., and seedlings become pale and stunted and die soon. The fungal pathogens invade the seedlings producing water-soaked lesions at the soil level, reddish brown sunken lesions and girdling the hypocotyl, and the seedling may collapse, on infected seedlings, dark lesions may expose to the stem and roots. *Rhizoctonia bataticola* fungus causes three types symptoms viz., seedling diseases, sore-shin and root-rot. One to two weeks old seedlings is attacked by the fungus at the hypocotyl and causing black lesions, griddling of stem and death of the seedling. The disease "sore shin" is known as damage caused by hypocotyl in the United States (Atkinson 1892). Sore-shin diseases occur on 4 to 6 weeks old plants, dark reddish-brown are formed on the stems near the soil surface and plant breaks at the collar region. The root rot symptom emerges usually at the time of maturity of the plants.

Causal organism

Fungal genera associated with seed deterioration are *Fusarium* (Klich 1986), *Rhizoctonia* (Brown and McCarter 1976), *Pythium* (Devay *et al.* 1982), and *Thielaviopsis* (King and Barker 1934) which cause damping-off disease and colonize the weak cotton plants. Rhizoctonia fungal hyphae are septate and produce black irregular sclerotia. The *Rhizoctonia bataticola* and *Rhizoctonia solani* fungi cause root rot of cotton, present in the soil with higher moisture.

Disease cycle

Rhizoctonia solani grows rapidly in soil under favourable conditions, the fungus survives in soil/plant debris as sclerotia. The spread of sclerotia by irrigation water, implements, heavy wind and cultural operations. The disease is mainly soil-borne and the pathogen can survive in the boil as sclerotia. Dry weather following heavy rains, high soil temperature (35-39° C), and low soil moisture (15 -20 per cent) are favourable to the disease. *Pythium* spp. overwinter as oospores and infect through germ tube and encysted zoospore. *Fusarium* spp. and *Thielaviopsis basicola* overwinter as chlamydospores in soil and plant residues for years. The sowing of cotton seeds in sandy soils with low orgnic matter increases the susceptibility of cotton to fungal pathogens. Seedling diseases are prevalent in a cool and wet climate, deep planting, poor seedbed conditions, and nematode increase the problem.

Management

Cotton seedling disease control through preventive rather than curative treatments. Rotation with non-host monocotyledonous crops like wheat, corn, and sorghum is useful in reducing the inoculum. Planting in raised beds and improving soil drainage can help in controlling seedling diseases. Disease-resistant varieties and planting of good quality seeds are recommended. Eradication of infected plant parts helps to control the seedling diseases of cotton. Seed treatment before sowing with Thiram, Azoxystrobin, and Metalaxyl is effective against these diseases. Seed treatment with antagonists such as *Trichoderma viride* and *T. harianum*. The combination of fungicides of Carboxin+Thiram+Metalaxyl besides Fludioxonil+Metalaxyl is more effective against seedling root rot (Nemli and Sayar 2002). The bioagents reduce fungal population and inoculum level in the soil as commercially available such as Kodiak, Subtilex, and Deny are suggested against seedling diseases (McSpadden Gardener and Fravel 2002). The effects of *Pseudomonas fluorescens* and *Bacillus subtilis* bacteria against *Rhizoctonia solani, Colletotrichum gossypii*, and *Fusarium* spp. (Pleban *et al.* 1995 Erdoğan *et al.* 2016, Wang *et al.* 2004). Intercropping with moth bean and application of $ZnSO_4$ kg/ha where the disease is prevalent. Drench the with Carbendazim (0.1%).

2. Alternaria Leaf Spot

It is a common foliar disease cause leaf blight found almost in every cotton-growing worldwide. This disease was first reported in the United States (Atkinson 1892; Paulwetter 1918). Later, similar spots were found on cotton in Nigeria (Jones 1928) and India (Rane and Patel 1956). The disease occurs in all stages but is more severe when plants at 45-60 days old.

Symptoms

The first symptoms appear as small, circular brown, grey-brown to tan lesions with purple margins on leaves. These spots vary from 1 to 10 mm in diameter in concentric zonation on older leaves. As the disease advance and lesions coalesce, become irregular and necrotic. The affected leaves become blighted, brittle, and crack showing a shot hole. The disease severe on lower leaves than on upper leaves and affected premature defoliation. Under humid weather conditions, sporulation of the fungus results in black sooty masses on necrotic lesions and also appears on the stem, bracts, and bolls.

Causal organism

Diseases caused by *Alternaria macrospora* Zimm. Earlier, it was recognized as *A. alternata* (Fr.) Keissler in Egypt and Russia (Kamel *et al.* 1971; Dzhamalov 1973), but recent reports from Zimbabwe described it as *A. macrospora* (Hillocks 1991). *A. macrospora* content cylindrical to slightly tapering conidiophores, they are septate, erect, flexuous, and pale brown. The conidia are produced singly or in chains light to dark brown with 4–9 transverse septa and several longitudinal septa, ellipsoidal, melanized, and obclavate to obpyriform with narrow beak (Ellis 1971). There is variation in conidial size (Ellis 1971; Sangeetha and Ashtaputre 2015; Venkatesh and Darvin 2016; Waghunde *et al.* 2018).

Disease cycle

The pathogen survives in the dormant mycelium of infected crop debris. The secondary spread is mainly by air-borne conidia. The infected seeds are the main cause of inoculums and infected seedlings support the early stages of an epidemic. Potassium-deficient soils favour the development of disease (Hillocks 1991). Air currents and water splashes spread conidia onto healthy plants. Wet and humid weather and temperature (about 27⁰C) favour disease development; the optimal temperature for disease development is between 20 and 30 C, and cotyledons to the lower leaves initiate the disease. Pathogen kills leaf tissues and produces abundant spores on the surface of the lesions within a few days under favourable conditions. Maximum sporulation of *A. macrospora* on susceptible varieties and defoliation of the leaves (Bashi *et al.* 1983), responsible for bolls damaged and seed infection (Bashan 1984). Symptom develops by physiological stress to plants like premature senescence. The shedding of leaves of infected plants and infected seed planting complete the disease cycle. High humidity and intermittent rain temperature of 25- 28⁰ C favours the disease incidence.

Management

Pathogens survive on infested crop debris in the soil so removing and destroying the infected plant residues, deep summer ploughing, and avoiding seeds from infected crops reduce inoculum production in the field. Crop rotation with cereals may reduce seedling infection. Application of potassium fertilizer to maintain the soil fertility level (Hillocks 1991). Cultivation of resistant varieties and planting of healthy seeds are recommended. Higher phenol contents might be responsible for resistance to the disease. Foliar application of fungicides is economically useful, spray Mancozeb@ 0.25% or Copper oxych loride @0.3% at the initiation of the disease at 15 days interval four to five sprays help in reducing the primary inoculums. Seed treatment with broad-spectrum fungicides like Strobilurins (Trifloxystrobin) and sterol biosynthesis inhibitors (Ipconazole) is effective in protecting the cotyledons of emerging seedlings. Seed treatment with bioagent *Pseudomonas fluorescens* at 10 g/kg seeds may reduce the disease intensity.

3. Grey mildew/Areolate mildew

Grey mildew was first reported in the United States (Atkinson 1891). In India it is known as grey mildew (dahiya) and false mildew/areolate mildew in the United States, white mildew in South America (Hillocks 1991). It is common in India, East Africa, and South America, little importance in the United States.

Symptoms

Symptoms first appear after first boll set on lower leaves as irregular-angular lesions and measuring 1–10 mm in diameter. Light green to yellow green translucent spots surrounded by veinlets (areolate) on upper surface, under surface of the leaves white mildew-like growth can be observed as a result of sporulation (conidia). Lesions may white on upper surface of the leaves under high humidity condition. Later, the lesions turn dark brown and become necrotic. On cotyledons circular water-soaked, reddish brown chlorotic spot can seen. In the severe infection premature boll opening and defoliation occur. In severe infections, leaves turn yellowish brown and fall off prematurely.

Causal organism

Disease grey mildew is also called as areolate mildew caused by *Ramularia areola* (Atk.) (synonym: *Ramularia gossypii* Speg.).The anamorph of fungus is *Cercospora gossypina* Cooke (Ehrlich and Wolf 1983), and teleomorph known as *Mycosphaerella areola* Earle (Ehrlich and Wolf 1983; Gouws *et al.* 2001). The conidiophores of *Ramularia areola* fungus is hyaline septate and conidia 14–30 x 4–5 μm. Spermogonia appear as black dots on the lower surface of fallen leaves the lesions. These perithecia produce fusiform asci having eight elongated, biseriate ascospores which (Ehrlich and Wolf 1983).

Disease cycle

The fungus has three phases during its life cycle. First stage is conidial stage as appears on the underside surface of the leaves, second stage is spermagonial that develops on fallen leaves and the third stage ascogenous produces on partially decayed leaves. Conidia and ascospores are the primary sources of inoculums and disseminated by wind and irrigation water. Conidia and ascospores required free moisture for germination with range of temperature 16–34⁰ C. Moist conditions with sporadic rains are favorable for development of the disease.

Management

Cultural practices such a destruction of crop residues, deep plowing and crop rotation can reduce the primary inoculums. Foliar application of Benomyl at 200–300 g/ha is effective in controlling grey mildew of cotton. Spraying twice with 1% Copper fungicide or dusting twice with Wettable Sulphur (0.2%) and Carbendazim (0.1%). The best method for control is using resistant cultivars.

4. Boll Rot Disease

Boll rot occurs most of cotton-growing regions in the world, result poor quality of lint produce. Yield losses in high humidity areas during late summer. Numerous microorganisms are associated with boll rots; organisms directly attack cotton bolls and enter through insect wounds. Almost hundred microorganisms have been isolated from rotted bolls (Hillocks 1991). Commonly fungus isolated *Fusarium* spp. from rotted bolls of cotton-growing area. *F. oxysporum, F. roseum, F. solani,* and *F. moniliforme* are mainly isolated from rotted bolls in America, (McCarter *et al.* 1970). *F. moniliforme, F. roseum, F. solani,* and *Colletotrichum* spp. are responsible boll rot in Africa (Follen and Goebel 1973), and in India, *F. equiseti* has been reported cause of boll rot (Sharma and Sandhu 1985). Recently, a complex of *F. incarnatum-equiseti* is reported boll rot in cotton (Chohan and Abid 2018). Boll first starts as appearance of water-soaked necrotic lesions on the margins of the bracts. During moist conditions, these lesions enlarge, and white to gray/pale pinkish fungal growth covers the infected bolls. In severe cases bolls may drop from plants. Boll rots can be prevent by adopt agronomic practices to avoid excessive application of nitrogen, maintain low humidity, crop canopy and burning of crop residues. Eliminating seed-borne infections through Acid delinting. Spray Copper oxychloride (0.25%) or Carbendazim (0.1%) along with the recommended insecticide.

5. Fusarium Wilt

Fusarium wilt is common disease of cotton crop causing significant crop loss. The fungus is free living and persist in soil as chlamydospores and in association with the roots of cotton as well as on the roots of weeds. Fusarium cause wilt, pathogen colonized in vascular system of plant. The pathogen invades the root of cotton and proliferates within xylem tissues, and eventually spreads throughout the plant. Vascular wilt fungi are soil-inhabiting pathogens, but these may grow on crop residues in the absence of host for long periods in the form of thick-walled resting structures. Fusarium sporulate and cause infection by blocking the vascular system of cotton plant under favorable conditions. Fusarium wilt of cotton was first reported by Atkinson (1892) from America, generally found around the world wherever cotton is grown. Its originated from Mexico and spread to South America, the United States, Egypt, West Indies, Italy, Africa, Greece, Zimbabwe, China, France, Russia, and India (Menlikiev 1962; Cook 1981; Hillocks 1992).

Symptoms

Wilting symptoms appear any time of plant growth, i.e., from seedling to maturity. The first symptoms appear on infected seedlings as vein darkening, yellowing, and shriveling of young cotyledons, later necrotic and shed, and eventually seedlings may wilt and die. In plants, symptoms appear as marginal yellowing of lower leaves. Symptoms first appear from the older leaves at the base, followed by younger ones towards. Leaves yellow and drooping; plants stunted, gradually wilt and die of the whole plant. Discoloration starts from margin and spread to midrib. Browning and blackening of vascular tissue is the other important symptom, black streaks may be seen extending upwards to the branches and downwards to lateral roots. Discoloured ring can be seen in the woody tissues of stem by transverse section. Typically characterized by patchy appearance of field. Partial wilting also occurs is affected plant and can be see only one portion of the plant other remaining free. The symptoms appear within 2 months after sowing, wilting and death of plants, fewer balls are very small and open prematurely.

Causal organism

The asexual ascomycetous fungus belonging to the class Hyphomycetes. It is recognized as *Fusarium oxysporum* f. sp. *vasinfectum* Atk. Syn. & Hans. Snyder and Hansen (1940) parasitic grouped forms into formae speciales of the strains based on host specificity. Mycelium of *F. oxysporum* f. sp. *vasinfectum* white to grayish initially or bluish purple in color and produces two types of conidia viz, microconidia and macroconidia. Microconidia are small, one to two-celled, and elliptical in shape measuring 5–20 x 2.2–3.5 μm. Macroconidia are multinucleate, usually three to five septate, fusiform, sickle-shaped, measuring 27–48 x 2.5–4.5 μm. Resting spores are called chlamydospores mostly spherical, single or in chains, terminal or intercalary measuring 7–13 μm in diameter. The distinguishing feature of *F. oxysporum* from other *Fusarium* species is formation of chlamydospores with short conidiophores.

Disease cycle

Pathogens are soil borne and survive as chlamydospores in the absence of host and also survives as saprophyte on plant debris. Chlamydospores are sources of primary infection. Infection starts as conidia germinate on root surface, and mycelial mat covering the root surface. Later, the penetrating hyphae become systemic and multiply in the xylem vessels, and conidia are transported upward in transpiration stream and produce more mycelium. Cotton plants are about 5–6 weeks old usually appear wilt symptoms. Optimum temperature for disease develop between 20 and 27° C, and moisture contents of 80–90% favorable for disease.

Management

Wilt disease can be managed through using resistant varieties, cultural practices like mixed cropping, field sanitation, proper use of fertilizers and micronutrients, and crop rotation. Non-pathogenic bacteria like *Pseudomonas fluorescens* is found effective in reducing incidence of wilt. The diploid cottons (*G. arboreum, G. herbaceum*) are susceptible to disease. Grow resistant varieties; soil application of Carbendazim (0.2%) helps in managing the disease. Seed treatment with bio- agents *Trichoderma viride* 10 gm/kg, *Pseudomonas fluorescens* @ 10g/ kg seed, Thiram 75% WS 3g/ kg seed, Soil drenching with *Trichoderma viride* @ 5 kg/acre mixed with 200 kg moist FYM. Carbendazin 2 gm/l of water at the base of affected plants as well as surrounding healthy plants

6. Verticillium Wilt

Verticillium wilt pathogens can grow inside the vascular system, but these fungi grow outside the vascular tissues in the advanced stage of infection. Verticillium wilt may sporulate on plant residues after the death of the host with narrow host ranges. Verticillium wilt cvan survive for a long time in the absence of a host plant in the soil. Verticillium wilt of cotton was first reported in 1974, *Verticillium dahliae* was isolated from a few diseased plants of upland cotton growing in Arlington, Virginia. The first cotton-growing land wilt diseases also occurred in Australia (Evans 1967).

Symptoms

The infected young plants show yellowing, epinasty, and defoliation of the leaves, and in warm weather, they recover quickly and show stunted growth. The plant leaves show a mosaic pattern with yellow areas between veins and on margins; necrosis of leaves become dry, cross-section of infected stem shows discolouration of the vascular system.. Leaves and bolls of the plant remain defoliated and ultimately killed.

Causal organism

Five species of *Verticillium* are *V. albo-atrum* Reinke & Berthold 1879, V. dahliae Klebahn 1913, *V. nigrescens* Pethybridge 1919, *V. nubilum* Pethybridge 1919, and *V. tricorpus* Isaac 1953. Verticillium wilt disease is caused by *Verticillium dahlia* in cotton. Colonies on PDA of *Verticillium dahliae* grow moderately fast (2.0–3.5 mm at 20–25^{0}C). Microsclerotia are dark brown to black. Conidia are ellipsoidal to short cylindrical with erect and prostrate conidiophores.

Disease cycle

The fungus survives as dormant microsclerotia in the soil debris and soil depths of 40 cm. Microsclerotia colonies in response to exudates on the cotton root surface. Hyphae from colonies penetrate the xylem vessels through wounds. Hyphae surrounding the necrotic tissue of leaves, stems, and roots, microsclerotia are formed after a month and depending on moisture, dispersed in soil (Schnathorst 1981; Huisman and Gerik 1989; Bell 1992). Diseases progress through soil-borne sclerotia and seed-borne conidia as contact with healthy and diseased roots. Fungus grows well on simple sugars and amino acids that are normally found in root exudates and xylem sap.

Management

A combination of cultural, chemical, and biological methods can minimize the losses. Resistant cultivars control the wilt that can resist disease infection. Cultivate resistant varieties to reduce the inoculums and prevent the introduction and spread of the disease in soil. Use of fertilizer, control of soil moisture, planting time and tillage method, planting density, removal of weeds and crop residues, and solarization. Fungicides such as carbendazim can control seedling pathogens as well as prevent seed transmission of the pathogen (Shen 1985). Benzimidazole fungicides control the *Verticillium* wilt in glasshouse and field with different concentrations, 100 ppm of Benlate in water drenchers. *Trichoderma viride* has been used to control wilt (Fedorinchik 1964). In low organic matter soil, *Gliocladium roseum* may be better than *T. viride* as an antagonist of *V. dahliae* (Globus and Muromtsev 1990).

7. Anthracnose

Infected seedling;s symptoms appear as small reddish circular spots on the cotyledons and primary leaves. The lesions develop on the collar region, the stem may be girded led, causing seedlings to wilt and die. In mature plants, infection symptoms lead to stem splitting and shredding of bark. The most common symptom on the boll appears small water-soaked, circular, reddish-brown depressed spots. The lint is stained to the yellow- brown solid brittle mass of fiber. The infected bolls cease to grow and burst and dry up prematurely. Pathogen forms large numbers of acervuli on the infected parts, conidia are hyaline and falcate, borne single on the conidiophores. Numerous thick-walled setae in the black colour are produced in acervulus. The pathogen survives as dormant mycelium in the seed for about a year. The pathogen perpetuates the rotten boils and plants debris in the soil. The secondary spread is by airborne conidia. Prolonged rainfall at the time of boll formation and close planting influence the disease. Remove and burn the infected plant debris and bolts in the soil. Treat the delinted seeds with Carbendazim/ Carboxin@ 2g/kg, and Thiram at 4g/kg. Spray the crop at ball formation stage with Mancozeb @ 0.25%, Ziram @ 0.25% and Carbendazim @ 0.1% found effective.

8. Bacterial Blight (Angular leaf spot or Black arm)

Bacterial blight of cotton is causing significant losses in yield in the rainy season and most destructive disease (Delannoy *et al.* 2005). It was first reported in1891 in Alabama in the United States (Atkinson 1891). This disease causes 30% yield losses in various cotton-growing areas of the world (Ramapandu *et al.* 1979; Chidambaram and Kannan 1989). Yield losses generally between 10 - 30% in Asian countries, and up to 50% in African countries (Bayles and Verhalen 2007). The bacterium attacks all stages of plant growth from seed to harvest. In India, disease was first time observed in Tamil Nadu in 1918. The disease is of economic importance in Maharashtra,

Karnataka, A.P., Tamil Nadu and Madhya Pradesh. The disease reduces the quality of the cotton lint consequent loss of yield (Verma 1986). Bacterial blight starts with small water-soaked lesions on the leaves and seedlings and mature plants. Damage to the tissues gradually produces necrotic, angular, waxy, and marshy lesions on the leaf, these lesions are called bacterial blight, angular leaf spot, lesion of the black arm, and boll rot (Hillocks 1992).

Symptoms

The first symptom appeared as tiny lesions with water (dark green flaccid) and then spread to the bottom of young leaves (Verma 1986). Premature dropping off of the leaves is a common symptom. The bacteria infect the veins causing vein necrosis or vein blight. Black arm symptoms are noticed when bacteria infects branches and stem. Black arm infection causes weak of the stems (Innes 1983; Akello and Hillocks 2002). Severe infections of bacteria may also infect the bolls and cause boll rot. The lesions progress into characteristic angular shapes when leaf veins limit, lesions may appear on the upper surface of the leaf; wet or "greasy" lesions are often observed on the underside of the leaf. Typically five common phases of symptoms are noticed such as seedling blight- small, water-soaked, circular or irregular lesions develop on the cotyledons and spread to the stem through the petiole and cause withering and death of seedlings. Angular leaf spot- small dark green, water-soaked areas develop on the lower surface of leaves, gradually enlarge and become angular restricted by veins and veinlets, lesions become older and turn to reddish brown, and infection spreads to veins and veinlets, spots can see on both the surface of leaves. Vein blight/ black vein- infected veins cause blackening, veins and veinlets give a typical 'blighting' appearance, and bacterial oozes on the lower surface of the leaf, are formed crusts. The affected leaves become crinkled and twisted inward and withering. Black arm- on the stem and fruiting branches dark brown to black lesions are formed, which girdle the stem and branches to cause premature drooping off of the leaves, cracking and gummosis, resulting in the breaking of the stem, black twigs give a characteristic "black arm" symptom. Ball rot- on the bolts, water-soaked lesions appear and turn into dark black and sunken irregular spots and spreads to the entire boll and shedding occurs, mature bolts lead to premature bursting of bolts and the pathogen infects the seed and cause a reduction in size and viability of the seeds.

Pathogen and Disease spread

The bacterium *Xanthomonas axonopodis* pv. *malvacearum* is a short rod with a single polar flagellum, gram-negative, non-spore-forming, aerobic, capsule forming and produces yellow colonies in a culture medium. Bacteria survive in the field on debris from previously harvested crops, and their initial inoculums were seed born (Mohan 1983). The bacterium sticks to the leaf surface and enters the leaf *via* the stomata and wounds cause defoliation, swelling of the stem, black arm, breakage of the weakened trunk, and detachment of the bolls. Wind-driven rain and running water along with rain were the main ways of spreading this bacterium (Brown and Ware 1958). Dust with storms produced wounds in the plant tissues and then later caused infections in plants. Seeds, machines, insects, and animals are also responsible for the transmission of this pathogen (Thaxton and Zik 2001).

Disease cycle

The bacterium survives on infected dried plant debris in the soil for several years. The bacterium is seed-borne and remains in a slimy mass on the fuzz of the seed coat. It multiplies in the seed as sown and infects the seedling. The primary infection initiates through seed-borne bacterium. The secondary spread through wind, windblown rain splash, irrigation water, insects and implements. The bacterium enters through wounds caused by insects and natural openings. As the disease progresses, diseased leaves defoliate early (Ridgway *et al.* 1984), and the disease spreads along the veins of the host plant known as the bacterial veins (Verma 1986). Bacterial ooze stained cotton fiber in diseased bolls (Brown and Ware 1958). Fruit positions become vulnerable to lesions. The bacterium has overwintered on infected seeds and plant residues and could survive at least 22 months upon seed (Kirkpatrick and Rothrock 2001).

Epidemiology

This disease was more severe in sub-humid regions with wind, rainfall ranging from 25.4 to 76.2 mm, and dust in the growing season (Kirkpatrick and Rothrock 2001). Disease infestation was higher in high-humidity areas that favoured the growth and spread of pathogens (Voloudakis *et al.* 2006). Black arm cotton infections damage 35% of bolls. Soil temperature of 28°C high atmospheric temperature of 30°C, relative humidity of 85 per cent, delayed thinning; late irrigation and potassium deficiency in soil are favorable for the disease development.

Management

Remove and destroy the infected plant debris in the fields as soon as possible. Rogue out the volunteer cotton plants and weed hosts. Use growth regulators to activate the defense mechanism of the plant against pathogens and apply irrigation timely. Manage seed sanitation to overcome the diseases by using acids, copper compounds and chlorine derivatives, Delint the cotton seeds with concentrated sulphuric acid at 125ml/kg of seed. Sown-resistant varieties like HG-9, G-27, Sujatha, and CRH 71 avoid agronomic practices during wet conditions, follow crop rotation with non-host crops, and the addition of potash to the soil reduces disease incidence. Disease forecasting models are important to farmers. *Gossypium herbaceum* and *Gossypium arboreum* are almost immune. *G. bordodense, G. hirsutum, G. herbaceum* var *typicum* have considerable resistance. Treat the delinted seeds soak in 1000 ppm Streptomycin sulphate overnight, and treat the seed hot water at 52-60°C for 10-15 minutes. Secondary spread in the field was checked by spraying the crop using Streptocyclin (0.01%) and Copper oxychloride (0.25%), Streptomycin sulphate (Agrimycin 100) along with Copper oxyrhloride at 3 week interval.

9. Leaf curl

The cotton crop is threatening to leaf curl disease in the north zone. Leaf curl was first noticed in Nigeria in 1912 (Kirkpatrick 1931) and other African countries such as Sudan, Tanzania, Egypt, and Malawi along with the South. It is considerable damage to cotton fields in north India since, 1993. Leaf curls disease is caused by the "gemini virus" and transmitted by Whiteflies (*Bemisia tabaci*). The symptoms begin on the newly emerging leaves with darkening and thickening of leaves; the leaves either cup upwards or curl downwards. Later leaf shaped enations appear on the lower side of the leaves from the main vein, and plants remain stunted. Excessive shedding of buds and bolls occurs. Characteristic symptoms of affected plants show vein swelling, upward/downward leaf curling, and stunted plant growth along with the formation of cup-shaped leaf-like-outgrowth undersides of leaves known as enations. Early infection causes severe stunting of plants with high yield loss, late infection causes mild symptoms (Sattar *et al.* 2013). The disease is associated with begomovirus complexes, which consist of a monopartite begomovirus and satellite molecule called a beta-satellite, previously known as DNA β, infection with satellite, and a molecule called alpha-satellite, previously known as DNA. The whitefly-transmitted genus Begomovirus (family Geminiviridae) consists of small, single-stranded, circular DNA genomes encapsidated quasi-icosahedral particles. All geminiviruses are monopartite, single genomics that is capable of replication, systemic movement, and infections. The begomoviruses are either bipartite or monopartite. New World Begomoviruses have bipartite genomes DNA A and DNA B, both are essential for successful infection (Stanley 1983). In the Old World bipartite begomoviruses cause disease in field crops, and large numbers of diseases are caused by monopartite begomoviruses. Single begomovirus *Cotton leaf curls Gezira virus* has been identified with Cotton leaf curl disease in Africa (Idris and Brown 2002).

Cotton leaf curl disease has been identified in numerous plant species such as cotton, hollyhock, okra, and Sida spp. (Tahir *et al*. 2011). *Cotton leaf curls Gezira virus* is a geographically widespread begomovirus and identified from different host plants including cotton from diverse areas in Asia (Tahir *et al*. 2011; Khan *et al*. 2012; Idris *et al*. 2014). Cotton leaf curl geminivirus causes a major disease of cotton in Asia and Africa. Leaves of infected cotton curl upward and bear leaf-like enations on the underside along with vein thickening. Early infected plants are stunted and yields reduce drastically.

The begomoviral disease can be managed by controlling the insect vectors whitefly using pesticides. Removal of alternate hosts like *Hibiscus esculentus*, and *Zinnia* sp, in reducing the diseases spread by whitefly from infected plants. The expression of insecticidal proteins and RNAi against whitefly has been demonstrated (Shukla *et al*. 2016; Raza *et al*. 2016; Javaid *et al*. 2016). The ability of begomoviruses-associated satellites to cause mixed infections is obstructed by engineering dual Begomovirus-*Bemisia tabaci* resistance in plants (Zaidi *et al*. 2017). Grow resistant varieties like HHH-223, H-1117, and H-1236, Sanitation, rogue out the infected plants, remove the entire reservoir and weed host, and spray 2-3 times with Monocrotophos @ 0.03% reduces the vector *Bemesia tabaci* population to check further spread of the disease. The *Gossypium arboreum* cotton is not affected by this disease.

References

Abbas Q and Ahmad S (2018) Effect of different sowing times and cultivars on cotton fiber quality under stable cotton-wheat cropping system in southern Punjab, Pakistan. *Pak J Life Soc Sci* 16:77–84

Akello B, Hillocks RJ (2002) Distribution and races of *Xanthomonas axonopodis* pv. *malvacearum* on cotton (*Gossypium hirsutum*) in Uganda. *J Phytopathol* 150:65–69

Amin A, Nasim W, Mubeen M, Ahmad A, Nadeem M, Urich P, Fahad S, Ahmad S, Wajid A, Tabassum F, Hammad HM, Sultana SR, Anwar S, Baloch SK, Wahid A, Wilkerson CJ, Hoogenboom G (2018) Simulated CSM-CROPGRO-cotton yield under projected future climate by SimCLIM for southern Punjab, Pakistan. *Agric Syst* 167:213–222

Atkinson CF (1892) Some disease of cotton: 3. Frenching. Bull Alabama Agric Exp Station 41:19–29

Atkinson GF (1891) Black rust of cotton: a preliminary note. *Bot Gaz* 16:61–65.

Bashan Y (1984) Transmission of *Alternaria macrospora* in the cotton seeds. *J Phytopathol* 110:110–118.

Bashan Y (1986) Phenols in cotton resistant seedling and susceptible to *Alternaria macrospora*. *J Phytopathol* 116:1–10.

Bashi E, Rotem J, Hans P, Kranz J (1983) Influence of controlled environment and age on development of *Alternaria macrospora* and on shedding of leaves in cotton. *Phytopathology* 73:1145–1147.

Bayles MB, Verhalen LM (2007) Bacterial blight reactions of sixty-one upland cotton cultivars. *J Cotton Sci* 11:40–51.

Bell AA (1992) Biology and ecology of *Verticillium dahliae*. In: Lyda SD (ed) Comparative pathology of Sclerotial-forming plant pathogens: a *Phymatotrichum omnivorum* symposium. Texas A & M University Press, College Station

Brown EA, McCarter SM (1976) Effect of seedling disease caused by the *Rhizoctonia solani* on subsequent growth and yield of cotton. *Phytopathology* 66:111–115.

Brown HB, Ware JO (1958) Cotton, 3rd edn. McGraw-Hill Book Company, Inc, New York, p 411.

Chidambaram P, Kannan A (1989) Grey mildew of cotton. Tech Bull Central Inst Cotton Res Regional Station Coimbatore, India.

Chohan S, Abid M (2018) First report of *Fusarium incarnatum-equiseti* species complex associated with boll rot of cotton in Pakistan. *Plant Dis* 103:151.

Cook RJ (1981) Fusarium disease in the People's republic of China. In: Nelson PE, Toussoun TA, Cook RJ (eds) *Fusarium* diseases: biology and control. Pennsylvania State University Press, University Park and London, pp 53–55.

Delannoy E, Lyon BR, Marmey P, Jalloul A, Daniel JF, Montillet JL, Essenberg M, Nicole M (2005) Resistance of cotton towards *Xanthomonas axonopodis* pv. *malvacearum*. *Annu Rev Phytopathol* 43:63–82.

Devay JE, Garber RH, Matherson D (1982) Role of *Pythium* species in the seedling disease complex of cotton in California. *Plant Dis* 66:151–154.

Dzhamalov A (1973) Irrigation and Alternaria leaf spot of cotton. *Zaschita Rastenii* 12:48.

Ehrlich J, Wolf FA (1983) Areolate mildew of cotton. *Phytopathology* 22:229–240.

Ellis MB (1971) Dematiaceous Hyphomycetes. CAB International, Wallingford, p 608.

Erdoğan O, Bölek Y, Göre ME (2016) Biological control of cotton seedling diseases by *fluorescent Pseudomonas* spp. *Tar Bil Der* 22(3):398–407.

Evans G (1967) *Verticillium* wilt of cotton the situation in the Namoi Valley. *Agric Gaz N S W* 78:581–583.

Fedorinchik NS (1964) Biological method of controlling plant diseases. *Vses Nauch -Issled Inst Zashch Rast Tr* 23:201–210.

Follen JC, Goebel S (1973) Rots of cotton capsules in irrigated crops in Côte d'Ivoire. Relationship with varietal characteristics, irrigation method and date of sowing. *Cott Trop Fibers* 28 (3):401–407.

Globus GA, Muromtsev GS (1990) The use of *Gliocladium roseum* as antagonist for defense of cotton from phytopathogenic fungi. In: Proceedings of the fifth international *verticillium* symposium, Leningrad, USSR, p. 90.

Gouws MA, Prinsloo GC, Van der Linde EJ (2001) First report of *Mycosphaerella areola*, teleomorph of *Ramulariopsis gossypii*, on cotton in South Africa. *Afr Plant Prot* 7(2):115–116.

Hillocks RJ (1991) Alternaria leaf spot of cotton with special reference to Zimbabwe. *Trop Pest Manag* 37(2):124–128.

Hillocks RJ (1992) Cotton diseases. Melksham UK, pp. 39–86 Hopkins JCE (1932) Some disease of cotton in southern Rhodesia. *Empire Cotton Growing Rev* 9:109–118.

Huisman OC, Gerik JS (1989) Dynamic of colonization of plant roots by *Verticillium dahlia* and other fungi. In: Tjamos EC, Beckman CH (eds) Vascular wilt disease of plants, NATO ASI (series H: cell biology), vol 28. Springer, Berlin, Heidelberg, pp 1–17.

Idris A, Al-Saleh M, Amer M, Abdalla O, Brown JK (2014) Introduction of cotton leaf curl Gezira virus into the United Arab Emirates. *Plant Dis* 98:1593.

Idris AM, Brown JK (2002) Molecular analysis of cotton leaf curl virus-Sudan reveals an evolutionary history of recombination. *Virus Genes* 24:249–256.

Innes NL (1983) Bacterial blight of cotton. Biol Rev 58:157–176 Iqbal Z, Sattar MN, Shafiq M (2016) CRISPR/Cas9: a tool to circumscribe cotton leaf curl disease. *Front Plant Sci* 7:475.

Javaid S, Amin I, Jander G, Mukhtar Z, Saeed NA, Mansoor S (2016) A transgenic approach to control hemipteran insects by expressing insecticidal genes under phloem-specific promoters. *Sci Rep* 6:34706.

Jones GH (1928) An *Alternaria* disease of the cotton plant. *Ann Bot* 42:935–947.

Kamel M, Ibrahim AN, Kamal SA, El-fahl AM (1971) Spore germination of Alternaria leaf spot disease. *U A R J Bot* 14:245–254.

Khan AJ, Akhtar S, Al-Shihi AA, Al-Hinai FM, Briddon RW (2012) Identification of cotton leaf curl Gezira virus in papaya in Oman. *Plant Dis* 96:1704.

King CJ, Barker HD (1934) An interval collar rot on cotton. *Phytopathology* 29:75.

Kirkpatrick TL, Rothrock CS (2001) Compendium of cotton diseases, 2nd edn. APS Press, St. Paul, MN, p 77.

Kirkpatrick TW (1931) Further studies on leaf-curl of cotton in the Sudan. *Bull Entomol Res* 22:323–363.

Klich MA (1986) Mycoflora of cotton seeds from southern USA. A three-year study of distribution and frequency. *Mycologia* 78:706–712.

Mccarter SM, Roncardi RW, Crawford JL (1970) Microorganisms associated with Aspergillus flavus boll rot in Georgia. *Plant Dis Resporter* 54:586–590.

McSpadden Gardener BB, Fravel DR (2002) Biological control of plant pathogens: research, commercialization, and application in the USA. Plant Health Prog. https://doi.org/10.1094/ PHP-2002-0510-01-RV.

Menlikiev NY (1962) *Fusarium* wilt of fine-staple cotton and a study of *Fusarium oxysporum* f. sp. *vasinfectum* strains as the causal agent of the disease in conditions of the Vakash Valley. *Rev Appl Mycol* 43:3381.

Mohan SK (1983) Seed transmission and epidemiology of *Xanthamonas campestris* pv. Malvacearum. *Seed Sci Technol* 11:569–571.

Nemli T, Sayar I (2002) Aydin Söke region the prevalence of cotton-depleting disease the factors and prevention of the possibility of investigation. Scientific and Technical Research Council of Turkey the TARP-2535 Ankara, p. 57.

Paulwetter RC (1918) The *Alternaria* leaf spot of cotton. *Phytopathology* 8:98–115.

Pleban S, Ingel F, Chet I (1995) Control of R. solani and S. rolfsii in the greenhouse using Endophytic bacillus spp. *Eur J Plant Pathol* 101(6):665–672.

Rahman MH, Ahmad A, Wang X, Wajid A, Nasim W, Hussain M, Ahmad B, Ahmad I, Ali Z, Ishaque W, Awais M, Shelia V, Ahmad S, Fahad S, Alam M, Ullah H, Hoogenboom G (2018) Multi-model projections of future climate and climate change impacts uncertainty assessment for cotton production in Pakistan. *Agric For Meteorol* 253-254:94–113.

Ramapandu S, Sitaramaiah K, Subbarao K, Prasada Rao MP (1979) Screening of cotton germplasm against bacterial blight caused by *Xanthomonas axonopodis* pv. *malvacearum. Indian Phytopathol* 32:486–487.

Rane MS, Patel MK (1956) Diseases of cotton in Bombay 1. Alternaria leaf spot. *Indian Phytopathol* 9:106–113.

Raza A, Malik HJ, Shafiq M, Amin I, Scheffler JA, Scheffler BE, Mansoor S (2016) RNA interference based approach to down regulate osmoregulators of whitefly (*Bemisia tabaci*): potential technology for the control of whitefly. *PLoS One* 11:e0153883.

Ridgway RL, Bell AA, Veech JA, Chandler JM (1984) Cotton protection practices in the USA and the world, pp. 265–365. In: Kohel RJ, Lewis CF (eds) Cotton. American Society of Agronomy/ Crop Science Society of America/Soil Science Society of America, Madison, pp 226–365.

Sangeetha KD, Ashtaputre SA (2015) Morphological and cultural variability in isolates of *Alternaria* sp. causing leaf blight of cotton. Karnataka J Agric Sci 28(2):214–219.

Sattar MN, Kvarnheden A, Saeed M, Briddon RW (2013) Cotton leaf curl disease– an emerging threat to cotton production worldwide. *J Gen Virol* 94:695–710.

Schnathorst WC (1981) Life cycle and epidemiology of *verticillium*. In: Mace ME, Bell AA, Beckman CH (eds) Fungal wilt disease of plants. Academic Press, New York, pp 81–111.

Sharma YR, Sandhu BS (1985) A new fungus associated with boll rot of *G. arboreum* cotton. *Curr Sci* 54:936.

Shen CY (1985) Integrated management of *Fusarium* and *Verticillium* wilt of cotton in China. *Crop Prot* 4:337–345.

Shukla AK, Upadhyay SK, Mishra M, Saurabh S, Singh R, Singh H, Thakur N, Rai P, Pandey P, Hans AL, Srivastava S, Rajapure V, Yadav SK, Singh MK, Kumar J, Chandrashekar K, Verma PC, Singh AP, Nair KN, Bhadauria S, Wahajuddin M, Singh S, Sharma S, Omkar URS, Ranade SA, Tuli PK, Singh PK (2016) Expression of an insecticidal fern protein in cotton protects against whitefly. *Nat Biotechnol* 34:1046–1051.

Snyder WC, Hansen HN (1940) The species concept in *Fusarium. Am J Bot* 27:64–67.

Stanley J (1983) Infectivity of the cloned geminivirus genome requires sequences from both DNAs. *Nature* 305:643–645.

Tahir MN, Amin I, Briddon RW, Mansoor S (2011) The merging of two dynasties - identification of an African cotton leaf curl disease-associated begomovirus with cotton in Pakistan. *PLoS One* 6: e20366.

Tariq M, Afzal MN, Muhammad D, Ahmad S, Shahzad AN, Kiran A, Wakeel A (2018) Relationship of tissue potassium content with yield and fiber quality components of Bt cotton as influenced by potassium application methods. *Field Crop Res* 229:37–43.

Tariq M, Yasmeen A, Ahmad S, Hussain N, Afzal MN, Hasanuzzaman M (2017) Shedding of fruiting structures in cotton: factors, compensation and prevention. *Trop Subtrop Agroecosyst* 20(2):251–262.

Thaxton PM, Zik KME (2001) Bacterial blight. In: Kirkpatrick TL, Rothrock CS (eds) Compendium of cotton diseases, 2nd edn. *American Phytopathological Society,* St. Paul, MN, pp 34–35.

Venkatesh I, Darvin G (2016) An overview on cotton *Alternaria* leaf spot and its management. *Int J Appl Bio Pharm Tech* 7(2):135.

Verma JP (1986) Bacterial blight of cotton. CRC Press, Boca Raton, FL, pp 278–279.

Voloudakis AE, Marmey P, Delannoy E, Jalloul A, Martinez C, Nicole M (2006) Molecular cloning and characterization of (*Gossypium hirsutum*) superoxide dismutase genes during cotton *Xanthomonas axonopodis* pv. *malvacearum* interaction. *Physiol Mol Plant Pathol* 68:119–127

Waghunde RR, Patel UT, Vahunia B (2018) Morphological and cultural variability of *Alternaria macrospora* causing leaf blight in cotton. *J Pharmacogn Phytochem* 7(3):3096–3099.

Wang C, Wang D, Zhou Q (2004) Colonization and persistence of a plant growth-promoting bacterium *Pseudomonas fluorescens* strain CS85 on roots of cotton seedlings. *Can J Microbiol* 50(7):475–481.

Zaidi SS, Briddon RW, Mansoor S (2017) Engineering dual begomovirus-Bemisia tabaci resistance in plants. *Trends Plant Sci* 22:6–8.

Printed in the United States
by Baker & Taylor Publisher Services